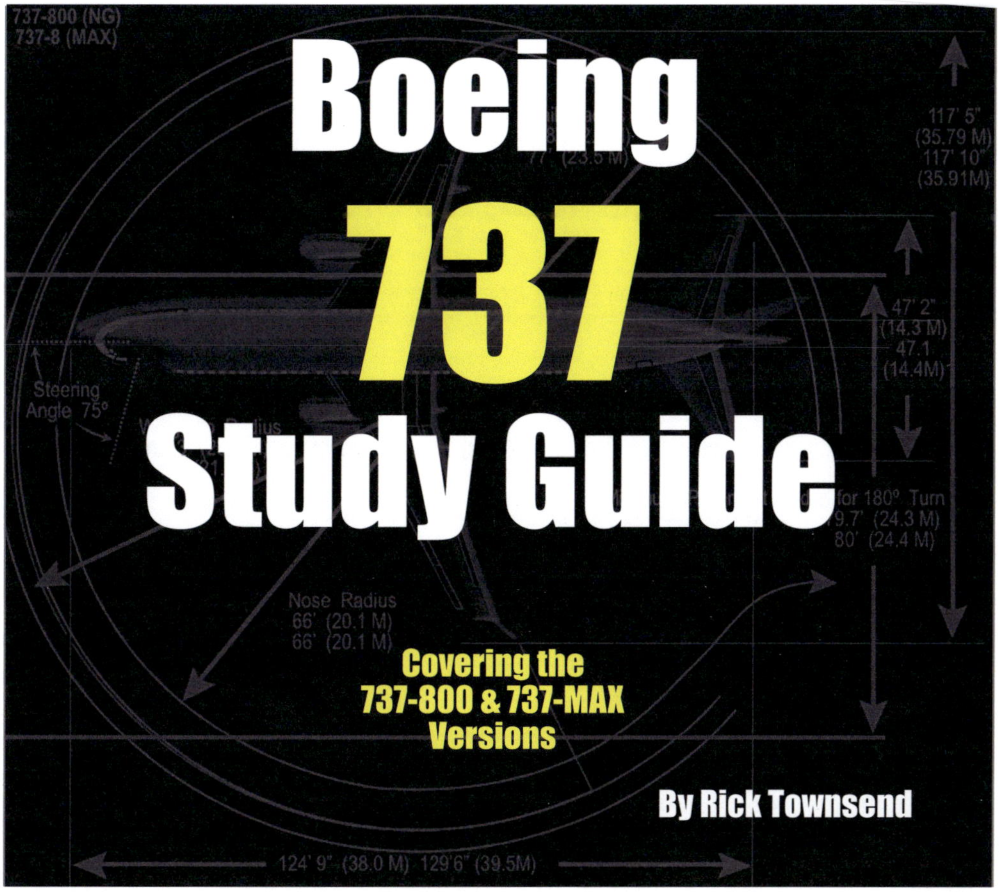

This book was originally developed for pilots at American Airlines. Neither American Airlines nor Boeing Company have endorsed or in any way been responsible for the content of this book.

As a courtesy to the pilots of American Airlines who have supported the development of these books, it is noted that this edition is current to the best ability of the author to make it so, and includes material relevant to:

Manual	Revision	Abbreviation
Aircraft Operating Manual	04.4	AOM
Aircraft Systems Manual	02	ASM
Quick Reference Handbook	12	QRH
Flight Operations Manual *	20*	FOM

* **Note**: *After Revision 20, FOM versions are password protected and cannot be updated for review by the author.*

Aircraft Model Designations

Model	Nose Numbers	Manual Designations
B737-800	3AA - 3PX	B737-NG
B737-8	3RA - 3VM	B737-MAX 8

Subfleets. B737-NG models are divided by three subfleets based on maintenance block designations for equipment variations.

- Block I are the first 76 aircraft (3AA-3DL) in the fleet. Major differences in the Block I are the BAE HUD, FMC software U11 and non-SFP.
- Block II and III aircraft (3DM-3LB) are non-SFP aircraft.
- Block IV are Short Field Performance (SFP) aircraft (3LC- 3PX).

Some systems and components are shared between the 737-800 and 737-8 models. Aircraft effectivity will be used to differentiate the aircraft.

See AOM Preface, Equipment Codes, p. 1-5

Revision Notes

January 01, 2023 Edition—(Initial issue for 2023 edition)
- Page 20 – Takeoff description: Added requirement to allow engines to stabilize on takeoff prior to engaging TO/GA
- Page 45 – APU Shutdown extended from 3 minute to 5 minutes on 3RA- 3VM aircraft.

March 24, 2023 Edition
- Page 6 – Clarified FOM reference for overweight landing considerations
- Page 20 – Updated Performance Manual reference for 27K Bump Thrust to AOM page reference.

June 5, 2023 Edition
- Page 11 – Removed Fuel Crossfeed Valve Operational Check from limitations section per OM revision.
- Page 20 – Updated Performance Manual reference for 27K Bump Thrust to AOM page reference.

August 14, 2023
- Page 5 – Added numbering to the Limitations Chapter for the convenience of those readers whose carrier uses section 18 numbering.
- Page 11 – Added non-Limitations Section material related to flight management restrictions and guidance.
- Page 11 – Added notes indicating removal of restrictions associated with ADs 2023-10-02, 2023-12-13 and 2023-12-11 related to radio altimetry restrictions and interference.
- Page 11 – Added Holding Speeds for ICAO as well as FAA airspace.
- Page 23 – Changed effective steering angle to 73° to reflect OM specification change.

November 3, 2023
- Page i – Note added to reflect password-protected manual revisions are not accessible or updatable by the author.
- Page 9 – Engine anti-ice restriction when not in actual or anticipated icing conditions added for aircraft 3RA - 3VM.
- Page 12 – Cabin Altitude Warning or Rapid Depressurization checklist: Steps 5 and 6 modified to match current operating manual nomenclature.
- Page 12 – APU Fire and Airspeed Unreliable Checklist reference page numbers in QRH updated.

Note: *After Revision 20, FOM versions are password protected. Until this is resolved the author cannot review or update manuals after this change was made. Use this guide at your discretion. Manufacturer or Air Carrier guidance always takes precedence.*

Published by Pilot Study Guides, LLC, Grapevine, Texas.
First Edition and Copyright 2009
Updated continuously (including copyright protection) through the current 2022 edition.

Graphics: With one exception, all graphics were produced by and are the copyright-protected property of the author, except where noted within this edition.

Acknowledgement: Thanks to Norebbo Design Studios, ***www.norebbo.com*** for allowing use of the profile drawing in the ***Airplane General*** section.

Corporation References: American Airlines™ and the Boeing Corporation™, referenced herein, have neither endorsed nor provided any proprietary information for use in these books. Their referencing is for informational awareness only.

ISBN: **978-1-946544-45-2** (Print)
ISBN: **978-1-946544-47-6** (Amazon Kindle)

Table of Contents

Introduction

What This Book IS

This Study Guide is a compilation of notes taken primarily from flight manuals, but also includes elements taken from class notes, computer-based training, and operational experience. It is intended for use by initial qualification crewmembers preparing for orals, and also for systems review prior to recurrent training or check rides. It is attempts to organize in one location all the buzz words, acronyms, and numbers the average pilot needs to know in order to get through the events above from an aircraft systems standpoint.

What this Book IS NOT

This series of books was originally developed for use by American Airlines pilots. It is for that purpose that operating manual references are for the manual sets for that airline as a courtesy to my loyal customers of many years. The books ***are*** ***not*** officially sanctioned in any way by any airline, however. The author assumes all responsibility for accuracy. (Forward corrections to the address provided, please!)

These books ***do not*** replace study of the operations manual, but instead provide a supplementary source of review material to complement study of official publications. Except for a few specific areas, it does not include regulatory guidance or other materials—just airplane systems details, some ideas for organizing the cockpit flow, cold weather operations, etc. The Study Guide is not printed on fancy paper or expensively bound. That makes it easy to tuck into your kit bag or briefcase and study as you travel. Therefore, for these same reasons, it is far less expensive than most commercially produced books of its kind. It is intended to be affordable and usable.

Suggested Uses

A good time to use this book is on layovers. A reasonable plan would be to attempt to review limitations, memorization Red Box items once per sequence, and the rest of the sections once per month. Reviewing sectional study outlines usually leads to trying to remember one of those great plumbing diagrams from the flight manual or synoptics, and so those also get reviewed in the process.

Limitations and Black Border items are laid out so that you may cover the answer with a 3x5 card and quiz yourself. Other sections are in outline form.

A word about acronyms—you will probably see most of the acronyms you have heard of before in this book, as well as a few new ones. Most of us avoid them when possible, using clues on aircraft instrument and systems management panels to jog memory. But for some topics, I've included all the ones I could. It is not necessary to learn all of them, but use the ones you need. Thankfully, the fleet management folks believe in keeping these "laundry lists" to a minimum, so you will see fewer of these than you will likely have seen in previous aircraft types.

♦ Unique Formatting Features

Throughout this book, several specific methods are used to highlight information. They include:

Recommended memorization sections

Material highlighted in a light blue box represents topics that are frequently discussed on orals or which simply need to be known without reference to manuals. This highlighting is found in systems notes, but not in the limitations section, which all requires a high level of retention. Please let me know of any recommendations you have for other information that should be included in this category.

System Annunciator Panel—Indicated by highlighted MC (Master Caution) and the specific word found on the respective L or R system annunciator panel, such as this example........ MC OVERHEAD

Panel or Warning Light Text—Text found on system panels or switches is indicated by bold, all capital letters, i.e. **LOW PRESSURE**.

Documentation Abbreviations

The following abbreviations are used in this guide:

Aircraft Operating Manual ....AOM
Aircraft Systems Manual ....ASM
Flight Operations Manual..FOM
Quick Reference Handbook ....QRH

__Note__: *These will be gradually implemented over several updates.*

Electronic Version

Digital versions have several special hyperlink features. Buttons at the top of each page represent functions available in the iBooks version of this study guide (iPad/smart phone), as well as the Kindle eBook version. To return to the Table of Contents or panel drawings tap the appropriate button at the top of each page. See web site for further details.

Table of Contents | Overhead Panel | Fwd Panel Ctr Console

The corner boxes on each page identify the book section and are not hyperlinked. INT

Panel drawing pages show graphics which are described in this book. Clicking on or touching the colored graphic panels in electronic versions of this book moves to the page with that graphic and its description.

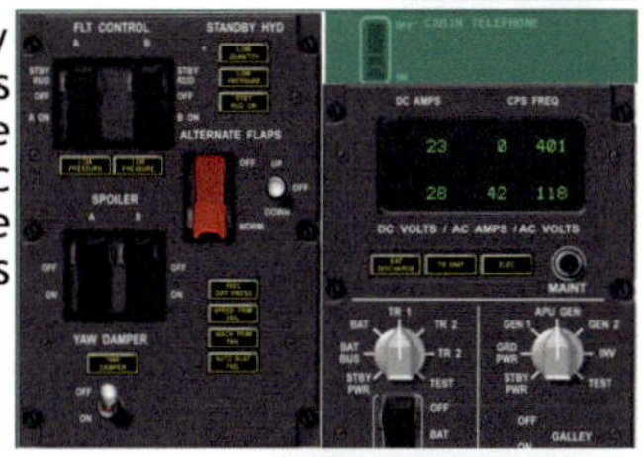

Green shaded areas are not depicted or discussed in this book.

Reproduction

Please don't photocopy this book. I have made a serious effort to keep the costs low for you. In fact, you can buy this book for about half the price of comparable material offered through other commercial publishers. There is a lot of work in this book, as you'll soon be able to see when you study the material. Pages with notes, when provided and annotated as such, may be freely copied by customers. Thanks for your cooperation.

Again, please send me any suggestions or corrections using the contact information below:

Rick Townsend ....Grapevine, Texas
....(817) 416-5533
E-Mail...rick@pilotstudyguide.com
Web Site.......................................www.pilotstudyguide.com

Revisions

Revisions are made on a regular basis. The downside of revisions is that updating your digital version causes you to lose highlighting and notes you have made. This is because each version includes a change to the printing date, so each page is updated on every new release. The methods for updating digital versions varies by platform. See the Pilot Study Guide web site for details.

To contact the Flight Academy Gift Shop and have a book mailed to your home, you may use the following:

Office 817-967-5540
Via Internet...............................www.crewoutfitters.com

Page References

Unless noted specifically with a "p." all references in this guide are to ***paragraph numbers***, i.e. for paragraph 2.3 on page 2-14, the reference is..2.3, p. 2-14

Publication Details

Compiled: January 23, 2025

Published by Pilot Study Guides, LLC, Grapevine, Texas

For ISBN and additional publication details, see title page i

Panels

Note: *Full color panels are those covered in this Study Guide. Green tinted areas generally do not have graphic reproductions or descriptions.*

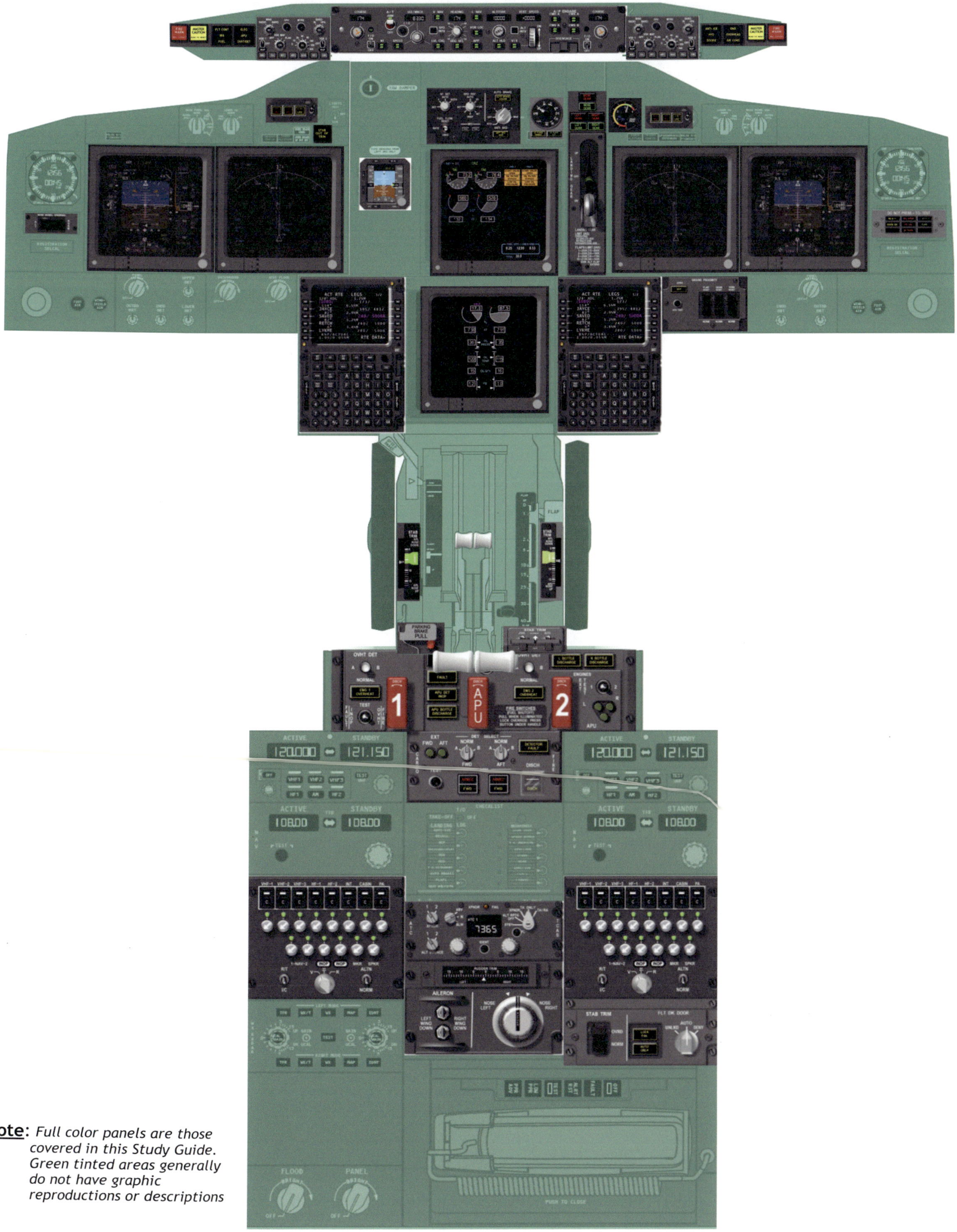

Note: *Full color panels are those covered in this Study Guide. Green tinted areas generally do not have graphic reproductions or descriptions*

Limitations

This Section is laid out for study with system areas on one side and the appropriate limitation on the other, so you can cover the answer with an index card and quiz yourself. Some of the limitations in Section 2 are not often encountered, such as alternate fuel limitations. For that reason, they have been shortened here for study. These will be listed in the left column. The Limitation column contains only a reminder to check the Ops Manual for the item in question. There are no memorization items, but all limitations need to be learned to a familiarity level.

Certain operational information and limitations must be memorized. Those items have ***are shown with titles and information bold italic print***. The AOM has these items in bold italics, and they are highlighted here with color as well. Memory items are those 1) which require timely action without reference to written material, 2) need only be memorized to the extent compliance is assured, but 3) exact wording is not required.

AFM items have been marked with a gray triangle (▲) to be consistent with the AOM formatting.

Items formatted with green background are from sections outside the AOM Limitations chapter. They are included in this section for ease of study.

18.1 General

General

Included Limitations: FAA, AA, and manufacturer's imposed warranty limitations

Not Included Limitations: Items that are
- Visibly identified on the panel / indicator
- Annunciated by a visual, tactile or aural warning
- Items whose compliance is assured by checklist accomplishment, MEL coverage, or placard.

Placarded Limitations: Generally provide actual certificate limitations, not warranty or AA imposed limitations

Disagreement Between Limitations: In event of disagreement between placard or instrument marking limitations and those shown in the AOM, more restrictive limitations apply.

Certification Status

737 models are certified: Transport Category, FAR Part 25 and FAR Part 36.

Configuration Required: Configuration must not vary from [what is] described in the AOM or QRH non-normal procedures, and data contained in the Flightdeck Forms Book (Yellow Book).

Altitude References: All references to altitude are Pressure Altitude, unless otherwise noted.

18.2 Operational Limits

Altitude Limits ▲

Maximum Operating Altitude: ±41,000'

Maximum TO & Landing Altitude: 8,400'

Maximum Operating Altitude with Flaps and/or slats extended: ***20,000 feet***

RVSM Altimeter Tolerances

Max difference — CA and FO Displays Inflight: 200 feet

On the Ground, Sea Level-5,000': 50 feet

5,001-10,000': 60 feet

CA or FO Displays and Field Elevation: 75 feet.

Standby Altimeter RVSM Accuracy: Does not meet accuracy requirements for RVSM airspace.

Speed Limits

Reference Speeds: All references to airspeed or Mach number relate to indicated airspeed or indicated Mach number, unless otherwise noted.

Speed Limitation Units: The airspeed limitations are given in both airspeed and Mach number. The limiting value is the lower of the two. Altitude determines which value will be lower.

Minimum Airspeed Limits: Intentional speed reduction below the onset of stall is prohibited unless a lower speed is specifically authorized for non-routine flight or training operations.

Maximum Airspeed Limits: Maximum operating limit speeds (VMO/MMO) shall not be deliberately exceeded in any regime of flight.

Maximum Operating Speed: 340 Knots ▲

Maximum operating Mach Number: .82 Mach (M_{MO}) ▲

Severe Turbulence Airspeed: ***280 Knots / .76 M*** ▲

Maximum Gear Extension Speed (V_{LO}): 270 knots / .82 Mach

Maximum Gear Extended Speed (V_{LE}): 320 knots / .82 Mach

Maximum Gear Retraction Speed (V_{LO}): 235 knots

Note: Observe V_{MO} pointer and gear/flap placards ▲

Design Maneuvering Speed

Design Maneuvering Speed (V_A): "A structural Design Maneuvering Speed, V_A, is defined for evaluating airplane structural design.

At or below this speed, Boeing airplanes are capable of sustaining a single input to any set of control surfaces (elevators, ailerons, rudder(s)) to their maximum available authority (as limited by control surface limiters, blowdown, or control stops).

These control surface inputs are to be in one axis (not in combination) and do not include control input reversal or oscillatory inputs. In addition, on Boeing airplanes at speeds above V_A, full rudder input is evaluated out to the maximum operating speed, VMO / MMO, and for some models, out to the design dive speed, V_D / M_D (where V_D / M_D is typically 30-60 knots /.05-.07 Mach higher than V_{MO} / M_{MO})."

Control Movements: However, even at speeds below V_A, transport category airplanes should be flown judiciously, avoiding full or abrupt control movements.

V_A Designation: Design Maneuvering Speed (V_A) data is shown on the V_{MO} / M_{MO} chart below, for the most critical gross weight. VA is not an identified speed on the PFD; Turbulence Penetration Speed may be used as a rough approximation for V_A.

V_A Chart:

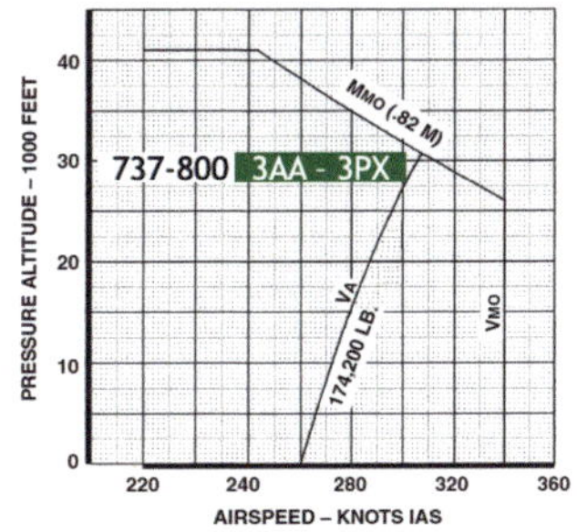

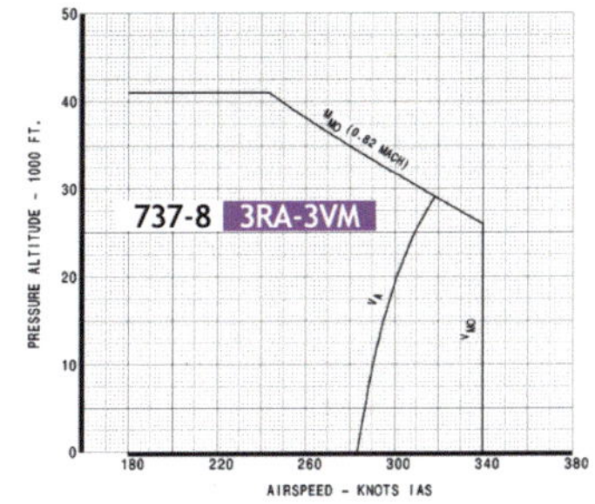

Maximum Flap Operating Speeds

1°:	250 Knots (240 Knots—AA Policy)
2°:	250 Knots (240 Knots—AA Policy)
5°:	250 Knots (240 Knots—AA Policy)
10°:	210 Knots
15°:	200 Knots
25°:	190 Knots
30°:	175 Knots
40°:	162 Knots 3AA - 3PX / 166 3RA - 3VM

Operational Limits

Operational Envelope ▲:

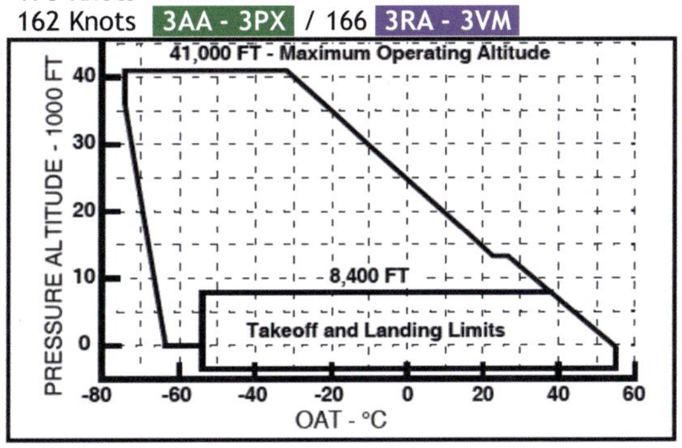

Weight Limits ▲

Maximum takeoff and landing weights may be further limited by runway analysis

Certificated Weights

	–NG 3AA - 3PX	–MAX 8 3RA - 3VM
Maximum Taxi:	174,700 Lbs	181,700 Lbs
Maximum Takeoff:	174,200 Lbs	181,200 Lbs
Maximum Landing:	146,300 Lbs	152,800 Lbs
Maximum Zero Fuel Weight:	138,300 Lbs	145,400 Lbs

Note: Refer to QRH> Overweight Landing procedure and FOM> Landing> Non-Normal/Emergency>Overweight Landing.

Wind Limits

Maximum takeoff and landing weights may be further limited by runway analysis

	3AA - 3PX	3RA - 3VM
Maximum Tailwind Component for Takeoff:	50	
Maximum Tailwind Component for Takeoff:	*15* [1]	
Maximum Tailwind Component for Landing:	*10* [1,2]	
Maximum Headwind Component for CAT II/III:	25	
HUD TO — Tailwind:	10	
Takeoff Crosswind Limits		
HUD TO:	10	
Dry:	33	32
Wet:	25	
Standing Water / Slush (>1/8 inch depth):	15	
Snow (Wet and Dry):	25	
Compacted Snow:	20	
Ice:	15	
Landing Crosswind Limits		
RCC PIREP		
6 Dry	33	32
5 Good	33	32
4 Good to Medium	33	32
3 Medium	25 [3]	
2 Medium to Poor	17 [3]	
1 Poor	15 [3]	
HUD SA Cat I / CAT II / III approaches:	15	
Visibility < 3/4 sm (RVR 4000) (1200 meters):	15	
Narrow Runway [< 148 feet (45 meters)]:	20	

[1] May be further limited by runway or visibility limitations.

[2] Authorized up to 15 knots if specified on the Ops Advisory page (##-7) for the airport or by F4 message and Ops Engineering has computed special data for a 15 knots tailwind.

[3] Reduce crosswind limits by 5 knots on wet or contaminated runways whenever asymmetric reverse thrust is used.

Ground Wind Limits

Crosswinds over 43 knots:	Limit thrust to a setting normally used for taxi
Crosswinds over 58 knots:	Limit engine thrust to idle, except when setting takeoff thrust on the runway
Door Wind Limits:	40 knots while operating, opening or closing entry or cargo doors

Miscellaneous Limits

Minimum Flight Crew: Captain and First Officer
Maximum Runway Slope ▲: ±2%

18.3 Aircraft General

Overwater Operations Seating Capacity

Infants and other children not occupying a seat: Not counted toward total

Notes: ***All aircraft have 172 passenger seats.*** The maximum aircraft capacity is not limited for all overwater flights requiring life rafts. The aircraft has 181 seats (172 passenger seats + 2 pilots' seat + 1W + 6 FA jumpseats). Life raft seating capacity is 193. This includes all required flight crew, passengers, observers and non-working crew members. Infants and other children not occupying a seat are not counted towards this limitation.

Passenger Evacuation

Prior to A/C Movement (Passengers on Board): At least one floor level exit must be open or armed

Escape Slide Retention Bar (Girt Bar): On revenue flights, must be inserted in the brackets on all doors during taxi, takeoff and landing.

Flight Deck Door & Access System ▲

Check Each Flight Day: Flightdeck door access system check must be accomplished prior to the First Aircraft Flight of the day.

Normal Access: Normal access to the flight deck will be accomplished as specified under Normal Procedures. Use of the access code is considered an emergency access request (except, prior to departure, when flight deck crew may be required to enter access code to gain entry to the flight deck).

Locking with POWER CUTOFF Switch OFF: If the DC Bus 2 is unpowered, place the **POWER CUTOFF** Switch in **OFF**, and use the deadbolt in the **LOCKED KEY INOPERABLE** position to lock the door.

Weather Radar — Weather Radar Ground Use

WARNING ***Inside a Hangar:*** *Avoid weather radar operations in a hangar*

Note **Test Mode in Hangar:** Above restriction does not apply

Personnel: Avoid weather radar operation when personnel are within the area normally enclosed by the aircraft nose radome. No personnel shall work directly with the antenna elements during periods of transmission.

Minimum Flight Crew Oxygen Pressure

Minimum Oxygen PSI for Dispatch		
Bottle Temp	Number of Crew	
°C:	2	3
50	530	735
30	495	690
10	460	645
-10	430	600

18.4 Air Systems

Air Conditioning—Cargo Compartment

Live Animal Cargo Carriage Location: Forward cargo compartment only

AC Packs Usage with Air Bleed Switch(es) ON: With either one or both engine bleed air switches ON, do not operate the air conditioning packs in **HIGH** for takeoff, approach or landing.

Note: The fire protection non-normal procedures take precedence over the statement regarding no air conditioning pack in HIGH during takeoff, approach, or landing.

Procedures requiring Pack Switch(es) HIGH: **CARGO FIRE**, **SMOKE/FUMES REMOVAL**, **VOLATILE LIQUID** — These checklists require the operating **PACK** switch(es) **HIGH** in order to increase ventilation for smoke removal.

Pressurization

Cabin Pressurization ▲

Maximum Cabin Pressure Differential: 9.1 psi (safety relief valve setting)

Unpressurized Flight — Maximum Altitude Following Inflight Depressurization: 14,000 feet. May be exceeded when terrain clearance requirements dictate.

Maximum Altitude when Dispatched for Unpressurized Flight: 10,000 feet

18.5 Anti-Ice, Rain

Icing Condition Definition ▲

Always: OAT/TAT 10°C (50°F) ***and***
Visible moisture in any form is present (clouds, fog with visibility one mile or less, rain, snow, sleet, or ice crystals.)

In Addition, During Ground Operations & Takeoff: When operating on ramps, taxiways, or runways where surface snow, ice, standing water, or slush may be ingested by the engines or freeze on the engines, nacelles

When Operating in Icing Conditions (as above): Engine ignition must be **ON**
WARNING Do not rely on airframe visual icing cues to turn engine anti-ice on. Delaying the use of engine anti-ice until ice buildup is visible from the cockpit may result in severe engine damage.

Anti-Ice Systems ▲

Engine Anti-Ice

Must be ON: During all ground & flight operations when icing conditions exist or are anticipated, except climb & cruise with temperature <-40°C SAT.
Before & during descent—all icing conditions, including temperatures <-40°C SAT

Engine Ignition: Must be ON when operating in icing conditions.

Use of Airframe Cues: WARNING Do not rely on airframe visual icing cues to turn engine anti-ice on. Delaying the use of engine anti-ice until ice buildup is visible from the cockpit may result in severe engine damage.

Maximum Temperature for Engine Anti-Ice: CAUTION : Do not use engine anti-ice when OAT (on the ground) or TAT (in flight) is above 10°C (50°F).

3RA - 3VM **Use of Engine Anti-Ice** when not in actual or anticipated icing conditions (AD 2023-15-05): CAUTION : Operation of EAI when not in actual or anticipated icing conditions may result in severe engine inlet damage or failure.

Wing Anti-Ice ▲

As Ground De-Icing: Do not use wing anti-ice as a substitute for ground de-icing / anti-icing and inspection procedures which are necessary to comply with operating rules.

Do Not Use Above Temperature Ground): Do not operate wing anti-ice on ground if OAT is above 10°C (50°F).

Ground or Flight: CAUTION : Do not use wing anti-ice when OAT (on the ground) or TAT (in

18.6 Autoflight

Radio Altimeter Restrictions ▲: See AOM 18.13.2, or page 11 of this book.

Minimum altitude for LVL CHG use, Final Approach: *1000' AFE*

Do not engage the Autopilot After Takeoff Below: *400' AGL*

Single Channel Operation Minimum Altitude on Approach ▲: *50' AGL*

Use of Aileron Trim with A/P Engaged ▲: *Prohibited*

Use of Dual Channel Autopilot for Approach: Prohibited

18.7 Auxiliary Power Unit (APU)

Starter Limit ▲ **APU Bleed Valve During Start:** May be open during engine start, but avoid engine power above idle

Operational Limits ▲

Bleed & Electrical: *10,000 feet*

Bleed Only: *17,000 feet*

Electrical Only: *41,000 feet*

APU Bleed Valve Must be Closed When:
- ✦Ground air is connected and Isolation Valve is open
- ✦Engine No. 1 Bleed Valve is Open
- ✦Isolation and Engine No. 2 Bleed Valves Open

APU Bleed Valve During Start: May be open during engine start, but avoid engine power above idle

18.8 Communications

HF Radio Use **Multiple Radio Selection:** If one HF radio is selected for transmission, deselect the other HF radio on all audio select panels to prevent audio interference.

During Refueling: CAUTION *: Do not operate HF radios during refueling operations.*

Broadband Antenna Hazard **RF Hazard:** Broadband system must be OFF during:
- ♦ Deicing
- ♦ Taxiing under a skybridge
- ♦ In a Hangar

CAUTION **Broadband Switch Location:** Passenger Oxygen Switch is in close proximity to broadband switch

18.9 Electrical

Circuit Breaker Resets Authorization: See 16.6 Troubleshooting/Circuit Breaker (CB) Reset. Pulling & resetting circuit breakers not authorized unless specified in a procedure.

18.10 Engines

Engine Ignition **Must be Used For:**
Takeoff (**AUTO** or **CONT**)
Landing (**AUTO** or **CONT**)
Anti-ice (**AUTO** or **CONT**)
Operation in heavy rain (**CONT**)

Engine Type
3AA - 3PX: CFM56-7B26K
3RA - 3VM: CFM LEAP-1B

Temperature and Time Limits

Engine Starter

	Starter Duty Cycle Limits	
3AA - 3PX:	2 Minutes ON	10 Seconds OFF
3RA - 3VM:	3 Minutes ON	10 Seconds OFF

Extended Engine Starter Motoring

	Extended Engine Motoring	
3AA - 3PX First Two Motorings:	15 Minutes ON	2 Minutes OFF
Third and Subsequent:	5 Minutes ON	10 Minutes Off
3RA - 3VM First Two Motorings:	5 Minutes ON	5 Minutes OFF
Third and Subsequent:		10 Minutes Off

Oil **Temperature for T/O Thrust** 3RA - 3VM: 31°C or above required before setting takeoff thrust

Approximate Warmup Time Prior to T/O Thrust: Approximate time for a cold soaked engine to reach 31°C oil temperature with engines at idle:

Ambient Temperature	Approximate Idle Time (Minutes)
0°C	3
-7°C	4
-18°C	6
-32°C	8
-40°C	9
-54°C	10

Quantity

	Minimum for Dispatch
3AA - 3PX:	12 Quarts

Reverse Thrust ▲

Where allowable:	Ground use only
In Flight Use:	*Intentional selection of reverse thrust in flight is prohibited.*
Use for backing:	Prohibited
Go-Around after Reverser Deployment on Landing:	Do not attempt

Engine Limit Display Markings ▲

	Indication
Maximum and Minimum Limits:	RED LINE
Caution Limits:	AMBER ARC
Normal Operating Range:	White Arc

Thrust **Antiskid Inoperative, Reduced T/O Thrust:** Operation with assumed temperature reduced takeoff thrust is not permitted with antiskid inoperative. See AOM 6p.4 SPAR Matrix.

EEC ▲ **T/O With EECs in Alternate mode:** Both EECs must be in Alternate mode
<u>Note</u>: *Understood to be "Hard" alternate mode per MEL* MEL Item 73.05
- ✦ 3AA-3PX 26K max (27K max, if planned) takeoff thrust must be used
- ✦ 3RA-3VM TO Takeoff Thrust must be used
- ✦ Do not use the FMC takeoff N1 or QRH V-Speeds
- ✦ Use of autothrottle for takeoff is prohibited.

27K Bump Thrust Use ▲ **3AA-3PX Only:**
- ✦ 27K operations authorized for use by dispatch and may only be selected if it is specified on the Load Agent generated TPS.
- ✦ 27K Thrust may not be selected or requested through ACARS if the TPS is based on any other thrust rating.
- ✦ KSNA operations are not part of this limitation.

3RA - 3VM: 14 Quarts

18.11 Flight Controls

Flaps — **Related Limits in Other Sections**

Altitude:	See page 6
Maximum Operating Speeds:	See page 7
Holding in Icing Conditions With Flaps Extended:	*Prohibited.*

Alternate Flap Duty Cycle

Minimum time after extending or retracting before subsequently selecting the switch again:	15 seconds
Cooling period after complete extend/retract cycle:	5 minutes
Maximum flap extension altitude ▲**:**	*20,000 feet* Also listed in the Operational Limit section, page 6

Flight Loads ***Large Flight Control Input Prohibition:*** *Avoid rapid and large alternating control inputs, especially in combination with large changes in pitch, roll, or yaw (e.g., large side slip angles) as they may result in structural failure at any speed, including below VA.*

Speedbrakes

Do Not Deploy Speed Brakes in Flight at Radio Altitudes Less Than ▲**:**	*1000 feet AFL*
In flight, do not extend the Speed Brake Lever beyond ▲**:**	The **FLIGHT** detent.
3RA - 3VM ***Speed Brake Lever beyond ARMED:***	*With Flaps 40 selected*
3AA-3PX **Use of speedbrakes between the down detent and the flight detent:**	Can result in rapid roll rates and normally should be avoided.
3RA-3VM **Use of speedbrakes in descent and roll response:**	Does not appreciably affect aircraft roll response.
While using the speedbrakes in descent:	Allow sufficient altitude and airspeed margin to level off smoothly.
After using speedbrakes in descent:	Lower the speedbrakes before adding thrust.
Speed Brake Use in Combination with Flaps General:	Simultaneous use should be avoided due to high sink rates
With Flaps 15 or Greater:	Speed brakes should be retracted ....AOM 9.2.2
3RA - 3VM ***With Flaps 40, Use of Speed Brakes:***	*Do not extend the Speed Brake Lever beyond the ARMED detent*

18.12 Flight Instruments, Displays

HUD

Use With Inop # 1 Radio Altimeter (RA)▲**:**	Prohibited
HUD Restrictions involving Radio Altimeter:	Use of head up display (HUD) with inoperative No. 1 radio altimeter (RA) is prohibited See AOM 18.13.2, or page 11 of this book.
Integrated Standby Flight Display (IFSD) — **Initialization:**	The Integrated Standby Flight Display (ISFD) must be initialized in accordance with approved procedures.
Display Electronic Unit (DEU) — **Software Errors:**	See notes on errors ....AOM 18.12.3, page 18-24

18.13 Flight Management, Navigation

Global Positioning System (GPS) **Updating** ▲**:** Must be disabled for approaches outside US National Airspace, if FMC database & charts not referenced to WGS-84 reference datum, unless other appropriate procedures are used.
Note: This limitation applies to operations to China and Russia. All B737 operations are referenced to WGS-84 datum.

Inertial Reference System (IRS)

Maximum Ground Alignment Latitude North:	78° 15 minutes
With IRS Magnetic Variation Greater than:5°:	5° **Note:** No B737 flights operate in these prohibited areas.
Maximum Flight Operations Latitude ▲ **N & S:**	82° N & 82° S (Dependent on A/C's ADIRU Magnetic Variation table data)
Exceptions For SEL 30 Aircraft:	Between 80W and 130W - 70N
Exceptions For SEL Non-30 Aircraft:	Between 80W and 170W — 73N / Between 120E and 160E — 60S

Note: SEL30 are aircraft with 2005 Magnetic Variation database.

Radio Altimeter Flight Restrictions ▲	
Radio Altimeter Flight Restrictions ▲: Source: AD 2023-10-02	All aircraft have been modified to be radio altimeter tolerant, therefore the restrictions contained in FAA AD 2023-10-02 do not apply.
-NG Added guidance for 5G C-Band interference—Approach, Takeoff, Landing and Go-Around ▲: Source: AD 2021-23-12	American Airlines has been approved to use an Alternative Method of Compliance for FAA AD 2023-12-13, therefore the restrictions contained in that AD do not apply.
-MAX Radio Altimeter 5G C-Band Interference—Takeoff and Landing Performance ▲: Source: AD 2023-12-11	American Airlines has been approved to use an Alternative Method of Compliance for FAA AD 2023-12-11, therefore the restrictions contained in that AD do not apply.
RNAV Use for SIDs and STARs — Building:	May not be built—must be loaded from current NAV data base. AOM 10.7.4 RNAV Visual and FMS/CVFP Visual Approaches; AOM 3.5.2 FMS Preflight, p. 3-19
Holding—Maximum Speed - FAA <6000':	200 KIAS — Reduce to holding speed 3 minutes prior to estimated initial time limit over holding fix.
>6000 through 14,000':	230 KIAS
>14,000':	265 KIAS — *May be restricted to 210 when on a low altitude chart*
	FOM Arrival > General > Holding Speed 8g.3.2
Holding—Maximum Speed - ICAO < 14,000':	230 KIAS — Exceeding maximum holding airspeed may result in aircraft excursions beyond the holding pattern protected airspace.
>14.000 through 20,000':	240 KIAS
20,000 through 34,000':	265 KIAS
>34,000':	0.83 Mach
	NOTE: Advise ATC immediately if unable to comply with the maximum holding airspeed and request an alternate clearance.
	Operational Requirements. For aircraft with an operational requirement (e.g., turbulence, icing, etc.), it may be necessary to hold at airspeeds in excess of the maximum speeds.
	Non-Radar Environment. In a non-radar environment, advise ATC you cannot accept the assigned hold.240 KIAS May be restricted to 210 when on a low altitude chart. FOM Arrival > General > Holding Speed 8g.3.3

Approach Categories

	Category	Approach Speeds (KIAS	
737 Aircraft:	D	141-165	AOM 10.1.3 Aircraft Approach Category

18.14 Fuel

Approved Fuels — **Standard:**	Jet A, Jet A-1
Alternate:	JP-6, JP-8
Prohibited:	JP-4, Jet B
Fuel Temperature	
Maximum — Fuel Tank Temperature:	49°
Minimum Tank Fuel Temp prior to T/O or inflight:	-43°C, or 3°C above fuel freezing temperature, whichever is higher
Use of Fuel System Icing Inhibitor — Effect on fuel tank temperature limit:	Does not change the minimum fuel tank temperature limit
Intentional running of center tank fuel pump with low level pressure light illuminated:	Prohibited.
Fuel Management — Tanks 1 & 2 Loading with Center Tank Containing over 1,000 Lbs:	Main tanks 1 and 2 must be full
Ballast Fuel — Authorization:	Ballast fuel is not authorized.
Intentional running of center tank fuel pump with low level pressure light illuminated:	Prohibited.
Fuel Imbalance — Allowable Lateral Imbalance	
Between Tanks 1 & 2 — Scheduled:	0 (Zero)
Maximum Imbalance:	Must not exceed 1000 lbs for taxi, takeoff, flight or landing.
Fuel Additive 3RA - 3VM Fuel Containing Kathon FP1.5: *(Applies to CFM LEAP-1B series engines)*	Use of biocide Kathon FP1.5 prohibited, 3RA - 3VM AD2020-14-09. **Note**: No crew action; fueling manual prohibits this additive
Minimum Wing Tank Fuel Quantity to Operate an Electric Motor-Driven *Hydraulic* Pump:	1675 Pounds ....ASM Hydraulics 13.20.2, p. 13.20.2, p. 13.20.5

18.15 Landing Gear

Speed Limits	See Operational Limits section, page 6
Brakes — When to Apply:	Do not apply brakes until after touchdown

18.16 Warning Systems

Ground Proximity Warning System (GPWS) — Use for Navigation:	Do not use the terrain display for navigation
ATC Deviation Authorization:	Deviation from ATC clearance only as needed to respond to EGPWS warning
Inhibiting EGPWS Enhanced Function:	EGPWS enhanced function should be inhibited (**TERR INHIBIT** set to **INHIBIT**) when aircraft position < 15 NM from airfield when: ◆ Operating from/to runways not incorporated in EGPWS database ◆ Specific approach procedures have previously been identified as potentially producing false terrain alerts **Note**: *For runways/airports not contained in the installed EGPWS terrain database, crews will be notified via F4 Pilot Alert Messages and/or Flight Plan/Dispatch Release remarks*....................................See also AOM 17.4

Emergencies

Recall items are in red and are accomplished from memory before the rest of the procedure is completed.

CABIN ALTITUDE WARNING or Rapid Depressurization

AIR 2.8

Condition: One or more of these occur:
- A cabin altitude exceedance
- In flight, the intermittent cabin altitude/configuration warning horn sounds and the CABIN ALTITUDE light illuminates.

1 Don oxygen masks and set regulators to 100%.
2 Establish crew communications.
3 Pressurization mode selector ....MAN
4 Outflow VALVE switch ....Hold in CLOSE
until outflow VALVE indication shows fully closed
5 **If** cabin altitude is **uncontrollable:**
FASTEN BELTS ....ON
PASS OXYGEN switch ....ON

▶▶ Go to the Emergency Descent Checklist on page 0.31

■ ■ ■ ■

6 **If** cabin altitude is **uncontrollable:**

— — — — Continue with QRC procedure. — — — —

APU FIRE — FIRE PROTECTION 8.2

Condition: Fire is detected in the APU.

1 APU fire switch ....Confirm .....Pull, rotate to the stop, and hold for 1 second
2 APU switch ....OFF

▶▶ Go to APU FIRE, QRH chapter 8 — Fire Protection 8.2

Runaway Stabilizer — FLIGHT CONTROLS 9.28

Condition: Uncommanded stabilizer trim movement occurs continuously or in a manner not appropriate for flight conditions.

1. Control column ....Hold firmly
2. Autopilot (if engaged) ....Disengage
3. Autothrottle (if engaged) ....Disengage
4. Control columnand thrust levers ....Control aircraft pitch attitude and airspeed
5. Main Electric Stabilizer trim ....Reduce control column forces
6. **If** the runaway **stops** after the autopilot is disengaged:
.....................Do **not** re-engage the autopilot or autothrottle.
7. If the runaway **continues** after the autopilot is disengaged:
STAB TRIM cutout switches (both)............ ....CUTOUT
If the runaway **continues**:
Stabilizer trim wheel ....Grasp and hold

▶▶ Go to Runaway Stabilizer, QRH Chapter 9

Airspeed Unreliable — FLIGHT INSTRUMENTS, DISPLAYS 10.1

Condition: Airspeed or Mach indications are suspected to be unreliable. (Items which might indicate unreliable airspeed are listed in the Additional Information section.)

Objective: To identify a reliable airspeed indication, if possible, or use the appropriate Flight With Unreliable Airspeed table on page 10.9 thru page 10.72 for the remainder of the flight.

1 Autopilot (if engaged) ....Disengage
2 Autothrottle (if engaged) ....Disengage
3 F/D switches (both) ....OFF
4 Set the following gear up pitch attitude and thrust:
Flaps extended................................ .....10° and 80% N1
Flaps up ....4° and 75% N1

▶▶ Go to Airspeed Unreliable, QRH Chapter 10

Emergency Equipment

16 FC / 144 MC (737-800NG, 3AA-3DL)
(Several variations – See AOM References below)

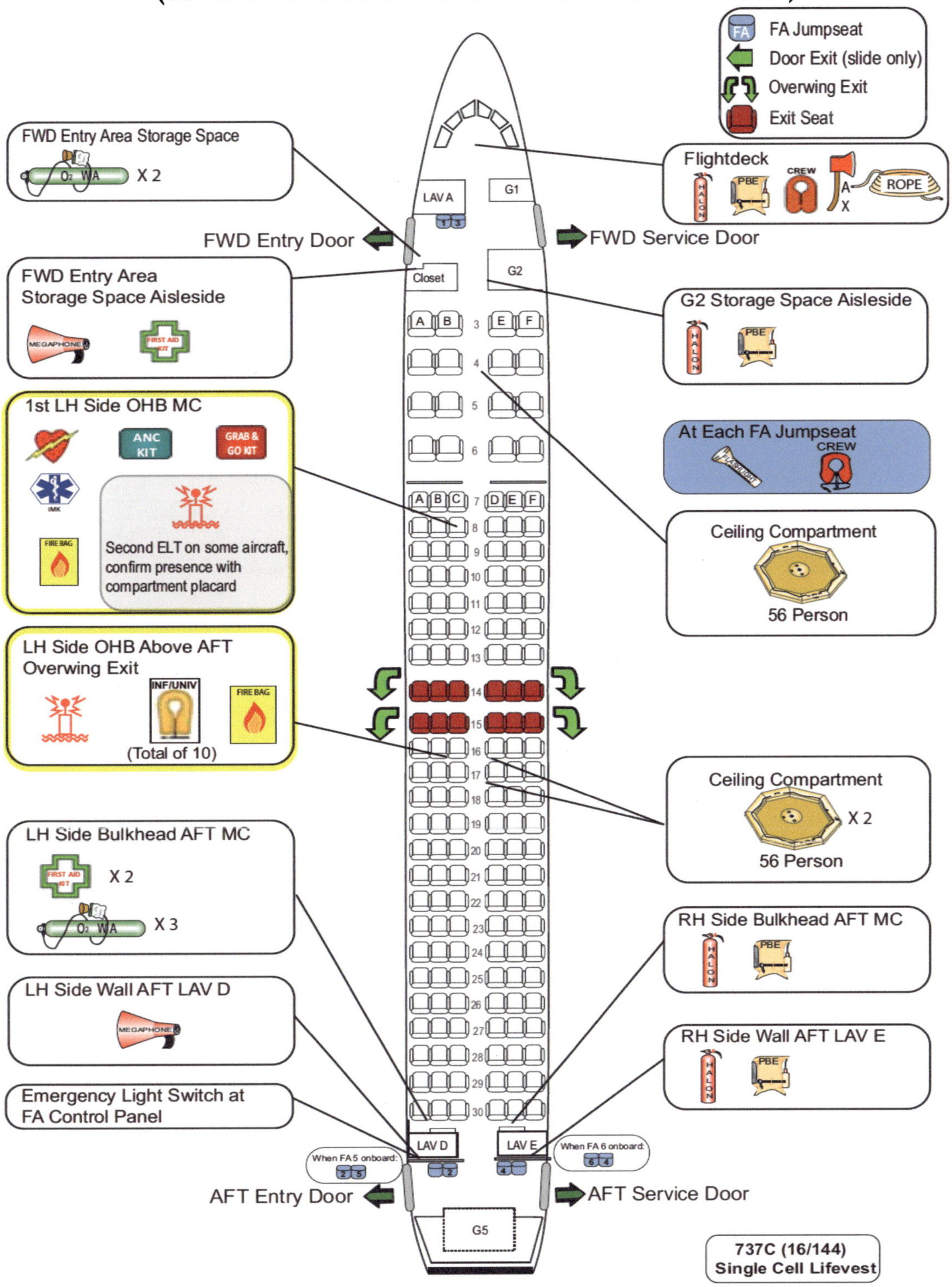

Aircraft Configuration

	Code	Business	Coach	Total
3AA-3PX	CE	16	144	160
3RA-3VM	45	16	156	172

AOM 20.3, p. 20.15 and following

Maneuvers

Notes: Only selected maneuvers are reproduced here for study. Notes are abbreviated for brevity and ease of study. Refer to QRH, MANEUVERS section for other procedures and full text of notes. Procedures are taken from the QRH, which states, **"This section is a consolidation of emergency maneuvers. Pilots are expected to be proficient in the performance of these maneuvers."**

References are to paragraph numbers, which are more easily found in the indices than page numbers.

Approach to Stall or Stall Recovery

All recoveries from approach to stall should be done as if an actual stall has occurred.

Immediately do the following at the first indication of stall (buffet or stick shaker).

WARNING

* If the control column does not provide the needed response, stabilizer trim may be necessary. Excessive use of pitch trim may aggravate the condition, or may result in loss of control or in high structural loads.

WARNING

** Excessive use of pitch trim or rudder may aggravate the condition, or may result in loss of control or in high structural loads.

Note

Do not use flight director commands during the recovery.

(AOM MANEUVERS 17.1):

Pilot Flying	Pilot Monitoring
First indication of stall (buffet or stick shaker)	
"My aircraft."	
Initiate the recovery: – Hold the control column firmly – Disengage autopilot – Disengage autothrottle – Smoothly apply nose down elevator to reduce the angle of attack until buffet or stick shaker stops. Nose down stabilizer trim may be needed.*	Monitor altitude and airspeed Verify all required actions have been done Call out – any omissions – any trend toward terrain contact.
Continue the recovery: – Roll in the shortest direction to wings level if needed** – Advance thrust levers as needed – Retract the speedbrakes – Do not change gear or flap configuration, except • During liftoff, if flaps are up, call for flaps 1	Monitor altitude and airspeed Verify all required actions have been done Call out – any omissions – any trend toward terrain contact. Set the FLAP lever as directed
Complete the recovery: – Check airspeed and adjust thrust as needed – Establish pitch attitude – Return to the desired flight path – Re-engage the autopilot and autothrottle if desired	Monitor altitude and airspeed Verify all required actions have been done Call out – any omissions – any trend toward terrain contact.

Upset Recovery
Nose High Recovery Actions & Callouts

WARNING

* Excessive use of pitch trim or rudder can aggravate an upset, result in loss of control, or result in high structural loads.

Note

Higher than normal control forces may be required to control the aircraft attitude when recovering from upset conditions. Be prepared to use a firm and continuous force on the control column and control wheel to complete the recovery.

(AOM MANEUVERS 17.2.1)

Pilot Flying	Pilot Monitoring
Nose high upset	
"My aircraft."	
– Disengage autopilot – Disengage autothrottle – Recover: • Apply nose down elevator. Apply as much elevator as needed to obtain a nose down pitch rate • Apply appropriate nose down stabilizer trim* • Reduce thrust • Roll (adjust bank angle) to obtain a nose down pitch rate* – Complete the recovery: • When approaching the horizon, roll to wings level • Check airspeed and adjust thrust • Establish pitch attitude	Call out – attitude, airspeed and altitude throughout the recovery – any continued deviation
	Verify all required actions have been completed

Upset Recovery
Nose Low Recovery Actions & Callouts

WARNING

* Excessive use of pitch trim or rudder can aggravate an upset, result in loss of control, or result in high structural loads.

Note

Higher than normal control forces may be required to control the aircraft attitude when recovering from upset conditions. Be prepared to use a firm and continuous force on the control column and control wheel to complete the recovery.

(AOM MANEUVERS 17.2.2)

Pilot Flying	Pilot Monitoring
Nose low upset	
– Disengage autopilot – Disengage autothrottle – Recover: • Recover from stall, if needed • Roll in the shortest direction to wings level. If bank angle is more than 90 degrees, unload and roll* – Complete the recovery: • Apply nose up elevator • Apply nose up trim, if needed* • Adjust thrust and drag, if needed	Call out – attitude, airspeed and altitude throughout the recovery – any continued deviation
	Verify all required actions have been completed

Windshear

Windshear Escape Actions & Callouts

Pitch Attitude. Aft control column force increases as the airspeed decreases. In all cases, the pitch attitude that results in intermittent stick shaker or initial buffet is the upper pitch attitude limit. Flight at intermittent stick shaker may be required to obtain positive terrain separation. Smooth, steady control will avoid a pitch attitude overshoot and stall.

Pilot Flying	Pilot Monitoring
When encountering a windshear	
"Escape."	
"My Aircraft"	
— Disengage autopilot — Push either TO/GA switch — Aggressively apply full thrust[1] — Disengage the autothrottle — Simultaneously roll wings level and rotate toward an initial pitch attitude of 15° — Retract speed brakes — Follow flight director TO/GA guidance (if available) without exceeding pitch limit indication (PLI)	Verify maximum thrust[1] Verify all needed actions have been completed Call Out: — Any omissions
— Do not change flap or gear configuration until wind shear is no longer a factor — Monitor vertical speed and altitude — Do not attempt to regain lost airspeed until windshear is no longer a factor	Monitor vertical speed and altitude Call Out: — any trend toward terrain contact, descending flight path, or significant airspeed changes
After Escape is successful	
— Resume normal flight — Retract gear and flaps as required	Issue PIREP to ATC

[1] Maximum thrust can be obtained by advancing the thrust levers full forward if the EECs are in the normal mode. If terrain contact is imminent, advance thrust levers full forward.

(AOM MANEUVERS 17.3.1)

Windshear Alerts: During Takeoff

Alert	Prior to V1	At or Above V1
Caution[1] "Monitor radar display."	Delay/reject the takeoff	— Maneuver as required to avoid windshear — Consider using maximum thrust
Warning "Windshear ahead. Windshear ahead."[2] or "Windshear. Windshear. Windshear."[3]	Delay/reject the takeoff	Perform the Windshear Escape Maneuver
Unacceptable Flight Deviations	Reject the takeoff	— Perform the Windshear Escape Maneuver — At VR, rotate normally to 15° no later than 2000 feet runway remaining

[1]Inhibited from 80 knots to 400 feet RA.
[2]Inhibited from 100 knots to 50 feet RA.
[3]Inhibited until rotation.

(AOM MANEUVERS 17.3.3)

Windshear Alerts: During Approach

Alert/Aural	During Approach
Caution[1] "Monitor radar display."	Continue the approach if able to avoid windshear Otherwise, execute a normal go-around and maneuver as required to avoid the windshear
Warning[2] "Go around. Windshear ahead."	Perform either: — A normal go-around or — The Windshear Escape Maneuver
Warning "Windshear. Windshear. Windshear." Unacceptable Airspeed Deviations	Perform the Windshear Escape Maneuver

[1]Inhibited below 400 feet RA.
[2]Inhibited below 50 feet RA.

(AOM MANEUVERS 17.3.3)

Enhanced Ground Proximity CAUTION Actions & Callouts

(AOM MANEUVERS 17.4)

Pilot Flying	Pilot Monitoring
Ground proximity *caution* aural alert activated	
Correct the flight path, aircraft configuration, or airspeed. If on approach and the stabilized approach requirements cannot be met, execute a go-around.	

Ground Proximity WARNING Actions & Callouts

Accomplish the following maneuver for any of these conditions:

- activation of the **"PULL UP"** warning
- 3DM-3VM activation of the **"OBSTACLE OBSTACLE PULL UP"** warning
- activation of the **"TERRAIN TERRAIN PULL UP"** warning
- other situations resulting in unacceptable flight toward terrain

Note

Aft control column force increases as the airspeed decreases. In all cases, the pitch attitude that results in intermittent stick shaker or initial buffet is the upper pitch attitude limit. Flight at intermittent stick shaker may be required to obtain positive terrain separation. Smooth, steady control will avoid a pitch attitude overshoot and stall.

Do not use flight director commands.

* Maximum thrust can be obtained by advancing the thrust levers full forward if the EECs are in the normal mode. If terrain contact is imminent, advance thrust levers full forward.

If an obstacle or terrain caution occurs when flying under daylight VMC, and positive visual verification is made that no obstacle or terrain hazard exists, the alert may be regarded as cautionary and the approach may be continued.

(AOM MANEUVERS 17.4)

Pilot Flying	Pilot Monitoring
Ground proximity *warning* aural alert activated	
"My aircraft."	
– Disengage autopilot – Disengage autothrottle – Aggressively apply maximum* thrust – Simultaneously roll wings level and rotate to an initial pitch attitude of 20° – Retract speedbrakes – If terrain remains a threat, continue rotation up to the pitch limit indicator or stick shaker or initial buffet	– Verify maximum* thrust – Verify all required actions have been completed Call out – any omissions
– Do not change gear or flap configuration until terrain separation is assured – Monitor radio altimeter for sustained or increasing terrain separation – When clear of the terrain, slowly decrease pitch attitude and accelerate	– Monitor vertical speed and altitude (radio altitude for terrain clearance and barometric altitude for a minimum safe altitude) Call out – any trend toward terrain contact

Traffic Advisory (TA) Actions & Callouts

Note

Maneuvers based solely on a TA may result in reduced separation and are not recommended.

(AOM MANEUVERS 17.5.1)

Pilot Flying	Pilot Monitoring
TCAS traffic advisory (TA) occurs	
Look for traffic using traffic display as a guide	
Call out – any conflicting traffic	
If traffic is sighted, maneuver if needed	

Resolution Advisory (RA) Actions & Callouts - Not in Landing Config

WARNING

Do not follow a DESCEND (fly down) RA issued below 1000 feet AGL.

Comply with the RA vertical guidance and ATC lateral guidance if there is a conflict between the RA and air traffic control.

Once an RA has been issued, safe separation could be compromised if current vertical speed is changed, except as necessary to comply with the RA. This is because TCAS -to- TCAS coordination may be in progress with the intruder aircraft, and any change in vertical speed that does **not** comply with the RA may negate the effectiveness of the aircraft's compliance with the RA.

Note

See additional Operating Manual notes.

(AOM MANEUVERS 17.5.2)

Pilot Flying	Pilot Monitoring
TCAS resolution advisory (RA) occurs	
If maneuvering is required: – Disengage the autopilot – Disengage the autothrottle – Smoothly adjust pitch and thrust to satisfy the RA command – Follow the planned lateral flight path unless visual contact with the conflicting traffic requires other action	
Attempt to establish visual contact.	
Call out – any conflicting traffic	

Climb RA in Landing Configuration Actions & Callouts

Note

When responding to an RA, the aircraft should be maneuvered only as much as needed to satisfy the RA.

Other critical warnings such as windshear or EGPWS take precedence over an RA.

Captain	First Officer
TCAS resolution advisory (RA) occurs	
— Disengage the autopilot — Disengage the autothrottle — Advance thrust levers forward to ensure maximum thrust is attained and call out:	
"FLAPS 15."	
— Smoothly adjust pitch to satisfy the RA command — Follow the planned lateral flight path unless visual contact with the conflicting traffic requires other action	Verify maximum thrust set. Position flap lever to 15 detent
	Verify a positive rate of climb and call:
	"Positive Rate"
Verify a positive rate of climb is indicated and call:	
"Gear Up."	Set the landing gear lever to UP
Attempt to establish visual contact.	
Call out — any conflicting traffic	

(AOM MANEUVERS 17.5.1, p. 17-41)

Rejected Takeoff

After RTO Considerations

When stopped, Review the following considerations and complete as required:

- Completion of Non-Normal checklist (if appropriate) for conditions which caused the RTO
- Not setting the parking brake unless passenger evacuation is necessary
- Review Brake Cooling Schedule for brake cooling time and precautions
- The possibility of wheel fuse plugs melting
- Advising the ground crew of the hot brake hazard
- The need to clear the runway
- The requirement for remote parking
- Wind direction in case of fire, position fire on the downwind side.
- Alerting fire equipment

Prior to 80 knots. The takeoff should be rejected for any of the following:

- activation of the master caution system
- system failure(s)
- unusual noise or vibration
- tire failure
- abnormally slow acceleration
- takeoff configuration warning
- fire or fire warning
- engine failure
- predictive windshear caution or warning
- if a side window opens
- if the aircraft is unsafe or unable to fly

Above 80 knots and prior to V1. The takeoff should be rejected for any of the following:

- fire or fire warning
- engine failure
- predictive windshear warning
- if the aircraft is unsafe or unable to fly

Note

During a rejected takeoff below 90 knots, autobraking is not initiated and the AUTO BRAKE DISARM Light does not illuminate.

Captain	First Officer
The captain decides to reject the takeoff	
"Reject, my aircraft."	
	If aircraft control is transferred
	"Your aircraft."
Simultaneously : — Close the thrust levers — Disengage the autothrottle — Apply maximum manual wheel brakes or verify operation of RTO autobrakes • If RTO autobrakes is selected, monitor system performance and apply manual wheel brakes if the AUTO BRAKE DISARM light illuminates or deceleration is not adequate	Verify: — Thrust levers closed — Autothrottle disengaged — Maximum brakes applied
	Call out — any omissions
	If AUTO BRAKE DISARM lights illuminates during rollout
	"Autobrakes off."
Manually deploy SPEEDBRAKE lever	
	Verify SPEEDBRAKE lever deployed — Visually confirm
	"Deployed."
	If SPEEDBRAKES do not deploy or fail to remain deployed
	"No speedbrakes."
	After call out, if captain fails to deploy SPEEDBRAKE lever - manually deploy
Apply reverse thrust up to the maximum amount consistent with conditions	
	Verify reverse thrust applied — Visually confirm
	If there is no REV indication(s) or the indication(s) stays amber
	"No reverse No. 1." or **"No reverse No. 2."** or **"No reversers."**
Continue maximum braking until certain the aircraft will stop on the runway.	Call out — any omissions
At 80 knots	
	"80."
At 60 knots	
	"60."
When stopping is assured: — Initiate movement of the reverse thrust levers to reach the reverse idle detent before taxi speed — After the engines are at reverse idle, move the reverse thrust levers to full down	
	Communicate the rejected decision to the control tower as soon as practical
When the aircraft is stopped	
"This is the captain. Remain seated. Remain seated. Remain seated."	

(AOM Takeoff Procedures 6.4)
See considerations, AOM 6.4.2

<table>
<tr><th rowspan="12">Emergency Descent

Note
Rapid descents are normally made with the landing gear up.

(AOM MANEUVERS 17.6.1)</th><th>Pilot Flying</th><th>Pilot Monitoring</th></tr>
<tr><td colspan="2">Emergency descent required.</td></tr>
<tr><td colspan="2">"Emergency Descent"</td></tr>
<tr><td>"My Aircraft"</td><td></td></tr>
<tr><td>**Altitude**:
— Set MCP lower
• Select a lower altitude and after descent is established select appropriate altitude
— Descend to 10,000 ft or minimum safe altitude (whichever is higher)
— Continue descent to 10,000 ft when able
Airspeed:
— Select LVL CHG
— Close Thrust Levers (both)
• Reduce thrust to minimum or as needed for anti-ice
— Speedbrakes - FLIGHT DETENT
— Set Mmo/Vmo
• If structural integrity is in doubt limit speed as much possible and avoid high maneuvering loads
Heading:
— Select a turn or proceed straight ahead</td><td>— Notify ATC and request altimeter setting
— PA- "**Emergency descent - Use Oxygen - Fasten Seat Belts**"
— FASTEN BELTS - ON
— ENGINE START switches - CONT
— Check safe altitude (MEA, MOCA, MORA, MSA, depressurization cloud altitudes)
When cabin has descended below 10,000 feet: — Notify flight attendants</td></tr>
<tr><td colspan="2">2000 Feet above level off</td></tr>
<tr><td></td><td>**"2000 feet above level off"**</td></tr>
<tr><td colspan="2">1000 Feet above level off</td></tr>
<tr><td></td><td>**"1000 feet above level off"**</td></tr>
<tr><td>1000 feet prior to level off:
— Speedbrakes - Retract</td><td></td></tr>
</table>

Normal Procedures Notes

Notes for this section focus on numbers and key items which are frequently needed in the cockpit, (and possibly on orals and simulators!) and for which there is usually not time to refer to the book. An example is for engine hot start considerations. In some cases, the information has been assembled from various places in the AOM. In these cases, reference page numbers have been provided. Other than this type of item, no attempt has been made to summarize the normal procedures sections.

Note: *This chapter contains notes and references to some of the more commonly needed numbers and page numbers buried throughout the normal procedures, abnormal and systems sections of the AOM, and some information from the Flight Training Manual. To the extent practical, it is organized in the order these items are encountered in a normal flight profile.*

Ground Operations

Use of Standard Thrust Not Permitted

Maximum Thrust Required
(22K, 24K, 26K, TO2, TO2, or TO Allowed)

- MEL items that may affect normal or back-up modes of generators, hydraulics, steering and brakes,
- If actual takeoff weight (GW on Takeoff Page 1/2) exceeds Assumed Takeoff Weight (ATOW). However, a new TPS may be obtained which authorizes Standard Thrust at the takeoff weight.
- With MEL / CDL items containing a takeoff weight correction UNLESS takeoff data is automatically corrected by TPS (TPAS), or MEL/CDL item prohibits standard thrust.
- If Engine Anti-ice is used and the TPS THRUST/V-SPEED section does not indicate ANTI-ICE ON.
- Whenever takeoff data is used from the (Contaminated Runway Correction/MEL) CRC/MEL message for an MEL/CDL item.

The following require Maximum Takeoff Thrust at the 26K, 27K Bump, or TO rating:

- When using Improved Performance.
- When the runway is contaminated by standing water, slush, snow, or ice.
- If windshear is reported or expected, including windshear advisories. Use a takeoff flap setting of 5, 10 (-MAX 8 only), or 15 unless limited by obstacle clearance and / or climb limit.
- When airport ops advisory (##-7) page requires Maximum Thrust

27K Bump thrust always requires the use of max thrust; standard thrust is prohibited.

Note: *It is the Captain's responsibility to ensure that no condition exists that prohibits use of standard thrust even though it appears on the TPS.*

Flaps 25

For the -NG, takeoff data for flaps 25 will only be available for 26K or 27K Bump thrust. TPS will limit flaps 25 to Max Thrust only, no Standard Thrust.

For the -MAX 8, takeoff data may be available for all thrust ratings, and may be used with standard thrust.27K Bump Thrust.

27K Bump Thrust (other than SNA)

27K Bump thrust is only available for a limited number of takeoffs. Its usage is limited to pre-authorized routes, and only when 27K Bump Thrust is needed to carry the payload.

As **27K Bump thrust will be available for takeoffs in these situations, outside of these the crew must not** select it without dispatch and load agent concurrence.

AOM - Gen Ops 1p.2.4, p. 1p-9

MAX 8 5G C-Band interference Procedures

See notes in Supplementary Flight Procedures.....AOM 21j.1.9

Single Engine Taxi

Not authorized when:

- Excessive taxi thrust is required (jet blast)
- Ramps & taxiways slippery/Braking action less than good
- Icing conditions (unless no significant precip is occurring)
- APU inoperative, except when maneuvering to position for a crossbleed start

Other considerations:

- MEL items that may affect normal or back-up modes of generators, hydraulics, steering and brakes,
- Ramp or taxiway configuration requiring tight turns, uphill slopes, crossing active runways, and airport construction areas,
- Proximity of the gate to departure / arrival runway,
- Completion of checklists and duties in a timely and safe manner without the added distraction inherent in single engine taxi ....AOM 5.2.1

APU Use

Single engine taxi may be conducted without APU operating

The remaining engine may be started with a crossbleed start .... AOM 5.2.1

Crossbleed Start ....AOM 21f.1.2

Autobrakes

Must be armed (If operative) for any of the following:

- Runway length less than 7500'
  - Less than 6000'—Requires Autobrakes 3 or MAX
- RVR less than 4000 or visibility less than 3/4 mile
- Tailwind
- Braking conditions reported less than good
- Runway wet / contaminated—standing water/snow/slush/ice

AOM Arrival 9.1.1, p. 9.4

Recommended whenever:

The runway is limited
Using higher than normal approach speeds
Landing on slippery runways
Landing in gusty winds or crosswind .....AOM Arrival 9.1.1, p. 9-4

Taxi Parameters

Thrust—N1 Settings

Congested Ramp Areas..35%
Elsewhere on Ramps 40%
Taxiways..45%

Speeds

Normal maximum recommended speed ....20 knots
Minimum radius turns..............................10 knots or less
Nose wheel steering maximum ....20 knots on takeoff roll
Nose wheel tiller steering wheel maximum speed .....20 knots
AOM 12.1.4, p. 12-20

Takeoff & Climb

Takeoff Parameters (Selected)

Takeoff Roll

Slight forward control column pressure recommended
Thrust levers – advanced to just above idle
Verify – both engines stabilize at approximately 40% N_1
Engines should be stabilized momentarily before engaging TO/GA.
First Press of TO/GA...........TO/GA Pitch Line appears at -10˚

Note: *Advancing thrust levers while simultaneously pressing* **TO/GA** *may lead to invalid thrust setting or* **TO/GA** *not engaging properly*

20 Knots ....Ground Path Vector Appears
..........................Runway Remaining Calculations Begin
40-60 Knots.............................Rudder becomes effective
60 Knots ....GND ROLL must be active
.............Gap in HUD TO/GA pitch line is set to MCP heading
....TO/GA pitch line jumps from -10˚ to +10˚ (*HUD*) & +15˚(*PFD*)
80 Knots ....Takeoff N1 set
...........................PM announces "80 Knots", PF Verifies
84 Knots ....THR HLD annunciation

Note: *Takeoff into headwind > 20 knots may make THR HLD appear before autothrottle can make final adjustments*

Approaching V1.............Control column is moved to neutral
At VR ....Rotate

After Airborne

10' AGL ....TO/GA pitch target line starts up from 10°
....Moves at 2° per second to match PFD FD pitch
"Once airborne".............................HUD FD & HSI appear
....FD may be tracked instead of gap in TO/GA line
If engine fails ....HUD FD pitch follows engine out logic
Compiled from various OM references: AOM 30.10, 42.3, 43.3

Tail Strike

Flaps	Liftoff Attitude (Degrees)	Minimum Tailskid Clearance (inches)	Tailstrike Pitch Attitude
1	8.5	13	11
5	8	20	11
15	7.3	25	11
25	7.0	29	11

Tail Strike Prevention
Rotation at 2-3° per second
Gusty winds—may need to hesitate at 10° to allow liftoff
Early or rapid rotation may result in tail strike

Initial Climb

All Takeoffs
Optimal climb angle speed ....V2 +15
Target Climb speed ....V2 +15 to V2 +25
50' AGL...........................If LNAV armed, becomes active
Above 400'....................Set Roll Mode (HDG SEL or LNAV)
1000' AGL ....Set Speed (Flaps up maneuver speed)
1000' AGL (NADP 2) / 3000' AGL (NADP 1)..............Set VNAV
AOM Noise Abatement Departure Procedures (NADP) 45.2

RNAV Departures
Bank angle limit below 200' AGL ....8°
FMS may command up to 25°, or 30° to stay on track
If RNAV route does not start at end of runway, HDG SEL may be required to intercept track
Two dot displacement in RNAV equals RNP value on legs

Initial Climb Speed—FMC ECON or 300/.78 if ECON not available ....AOM 45.5.6

Initial Bank Angle Limit
If airspeed within amber band (minimum maneuvering speed for takeoff flaps on PFD)—15°....................AOM 42.6.7

Low Visibility Takeoff

Setup..AOM 43.2.1
MCP Heading and courses ....Set Runway Heading
(Use published runway heading, NOT ILS course.)
CDU HUD MENU Page ....DATA ENTERED and VERIFIED
VHF Nav Radios ....SET ILS FREQ
(Use of backcourse localizer for T/O not authorized)
LOC Ident / Course ....CHECK
Airplane must be receiving a valid localizer signal for ground roll guidance
HUD Mode ....NP

Static runup – Recommended ....AOM 43.3.2

Prior to brake release....................**GND ROLL** cue visible
Note: *Abort if not visible by 60 knots* ....AOM 43.3.2
Localizer symbol & course arrow .Approx. aligned with runway
Takeoff Roll........................Follow **GND ROLL** guidance

Takeoff Benchmarks
Ground Path Vector (BAE 2020) appears ....20 Knots
Runway Remaining starts calculating after ....20 Knots
Ground Roll Cue (HGS 4000) provides guidance ....40 knots
TO/GA Pitch line Appears at +10° ....60 Knots
Other takeoff notes apply, see above, this section

Deviation Callout
If localizer deviation noted—F/O callout is........*"Centerline"*
Captain determination to continue visually/abort..AOM 43.3.1

Runway Remaining Cue

Runway remaining based on FMC takeoff runway
If intersection takeoff is entered in CDU, that point is used

Distance starts counting down> 20 knots

Box around Runway Remaining........................3,000'
(*Red and white centerline Light start point*)

Boxed Runway Remaining flashes ....2,000'
(*Amber edge light start point*)

Runway Remaining Disappears.........................1,000'
ASM 3AA-3DL 10.42.2; Displays p. 10.42-7

Landing

Landing Flap Selection

Normal Flap Settings.........15, 30 or 40

Flaps 15 consideration factors
High density altitude airports
Climb limited landing weight exceeded for flaps 30 or 40 landing

Flaps 40 recommended when
Braking action less than good
Tailwind
Wet / Contaminated runway
Runway length ≤ 7500' (Autobrakes 3 or MAX recommended)
AOM 9.1.3

Conditions Requiring a Go-Around (ILS)

Anytime: Localizer or glideslope indicator shows full deflection (2 1/2 half dots or more and magenta diamond goes hollow).
Prior to DA: Any required aircraft or ground equipment fails.
At DA: Requirements to continue approach below DA (14 CFR Part 121.651) have not been met.
Below DA: Reduction in visual references occurs which prevents PF from safely continuing the approach and landing.
The Captain determines safe landing within touchdown zone cannot be assured ....AOM 51d.4.2

Approach Techniques

Initial Setup ....**WARM-V**
Weather requirements ....Check
Include crosswinds, visibility, published minimums
Approach ....Load Approach from Database
....Brief Approach
RNP ....As Required
MCP—When cleared for the approach..............LNAV / VNAV
Set FAF altitude when on intercept *or* a portion of approach
VOR Updating....Turn Off **<PROG> <NAV STATUS> <Page 2/2>**

Captain Callouts ....LOC / GS Intercept
....Passing FAF

Passing FAF Actions..**MAC-V**
Missed approach Altitude.............Set when 300' below MAP
Altimeters—Verify CA–F/O ....± 100'
Clearance—Tower ....Call Passing FAF
V/S.......................Set prior to MDA if end point not RWXX
Note: *This section compiled from instructor notes and various OM references; NOT procedural, but technique recommendations.*

Briefings and Restrictions

Flight Attendant Briefing—Prior to gate departure

Known delays
Maintenance—cabin or galley writeup and deferrals in AML
Ground—Short-taxi & safety demo considerations
Cabin Door Handles—Push Down (FMV1 Preflight 35.5.1)

Crew
Number of F/As onboard and minimum crew requirements
See paragraph 3g.10.2 Flight Attendants.
Coordination of flightdeck crew meal service, if applicable
Time for meals is left to the CA's discretion
Cooperation should result in eating meals at desired time
Need to minimize noise & activity near crew rest areas

Security
Details not included here for sensitivity concerns

Turbulence
- Weather/turbulence forecast for route: duration & severity
- Suspension of service at at flight attendant discretion
- Notification required action for unexpected turbulence
- Notify flightdeck of cabin situation & injuries if applicable
- If not seated at jumpseat with interphone, stay seated to avoid injury until instructed to resume duties by Captain

International
- Gen DECs/Customs Immigration forms (if required)
- N Atlantic flights north of 66°N over Greenland—request FA 1/purser review applicable FA Safety Manual information reference Supplemental O2 Unit ....FOM 3c.4.2

Flight Attendants—Emergency TEST Acronym

T—TYPE of emergency: General description of the emergency
E—EVACUATION: Yes or No
Include any special considerations which may affect the use of exits.
S—SIGNAL: the following PA:
"This is the Captain. Evacuate. Evacuate. Evacuate."
Followed by turning on Evacuation Command (as installed).
T—TIME to landing ....FOM, 1N.6.3 Evacuation
....QRH, MISCELLANEOUS 0.45

Emergency PA to Passengers

Nature of emergency
Time to landing
Passenger cooperation with Flight Attendants
Reassure passengers. ....QRH MISCELLANEOUS 0.27

Stabilized Approach Deviation Callouts

Airspeed...+10 / -5 knots*
Rate of Descent......................<2000'—More than 2000 fpm
.................................<1000'—More than 1000 fpm*
Glide Slope ....> ½ Dot Deviation
Localizer..............................> ⅓ Dot (non expanded) or
Diamond touching Rectangle (expanded)
VOR ....Over 2° (20.18)* or Over 1 Dot (15.7)*
RNP ....ND Message—**UNABLE REQD NAV PERF-RNP**
CAT II/III .....**TAILSTRIKE** (HGS 4000) / **TAILST** (BAE 2020) Warning
Touchdown ......First 1/3 of Runway Not to Exceed first 3000' *
* PNF directs GO AROUND if PF not correcting
Descent Rate: AOM 10.2.2
Landing Configuration & Speed: AOM 10.2.2

Restricted Captain Requirements

Applies if
CA has < 100 hours after IOE in current aircraft type
Required 100 hours PIC in aircraft type may be reduced not to exceed 50% by substituting one landing for one hour of PIC time in aircraft type.

Restrictions *unless* FAA Exemption 5549 apply
- Chart increases required visibility ....FOM 18.6.4, Table 18.1
- Alternate Airport - Published minima need not be increased, but lowest allowable landing minima is 300 feet MDA/DH and one mile visibility or 4500 RVR / 1400 meters

Practical Effect of Exemption 5549:
Recognizes inherent safety of autopilot coupled approaches
Allows approaches using
- CAT II procedures to no lower CAT published minima or
- CAT I procedures to published CAT I MDA, no lower than RVR 1800'

See detailed charts ....FOM 18.6.4 FAA Exemption 5549

Low Experience FO: <100 Hours (FAR 121.438)

When Applies—FO has <100 hours flight time as SIC in type aircraft being flown, ***and***
Captain (PIC) is not a qualified Check Airman,

PIC must make all takeoffs when:
All special requirement airports
Visibility in the latest WX report at or below 3/4 mile
RVR for the runway to be used is at or below 4000 feet
Runway has water, snow, slush, or similar conditions which may adversely affect A/C performance
Braking action is reported less than good
Crosswind component >15 kts
Windshear reported in the vicinity of the airport
Any other condition PIC determines prudent ....FOM 18.6.5
Note: *CAT II/III operations are authorized with a low time FO*

Landing Distance Assessment Not Required if:

Runway length ≥ 7,000' ***and*** all of the following apply:
- Field elevation ≤ 2,500'
- Braking action GOOD or better
- Landing weight ≤ structural limit weight
- Flaps planned for 30 or 40
- No Tail Wind
- No MEL/CDL or non-normal items affecting affecting landing performance exist.................................AOM 9.1.4

Autobrakes—During landing, use autobrakes as recommended see *page 20* of this guide.

Cold Weather & Anti-Icing Notes

Cold Weather Operations Ref ....AOM 15.8

Cold Weather Operations

In effect when potential icing conditions exist with OAT ≤50°F / 10°C
Upper wing surface inspection is required

Frost/Ice Allowable
No ice, snow or frost allowable on any of these:
- Wings
- Engine inlets,
- Control surfaces
- Other critical surfaces
- Pack inlet & exit doors, outflow valves, pressure relief valves
- Landing gear, doors, actuating mechanisms & steering components
- Pitot tubes, static ports and stall warning vanes
- APU Inlet and exhaust
- Engines
- Exhaust ducts
- Inlet ducts
- Thrust reversers
- Doors and probes
- Inlet cowling
- Fan—free to rotate, but frost permissible

Allowed—Up to 1/8"
Bottom of wing under fuel tank ***or***
Within CSFF marked area on upper wing surface
Frost outboard of fuel measuring stick 4 may indicate frost on top of the wing
Thin frost on the fuselage where surface features such as paint lines are still evident is also permitted

Deicing—process of removing frost, snow, or ice
CAUTION —Operate APU during deicing only if necessary due to fume ingestion.................................AOM 15b.3.2
Leave APU bleed closed for ≥ 1 minute after fluid application

Anti-icing—Protection for limited time against formation of frost, snow or ice

One Step Deicing
Aircraft contaminated but icing conditions no longer exist—no further ice buildup expected
Allowable fluids
Heated Type 1
Outside US—Type II or IV
Hot water allowable if:
OAT over 40°F / 5°C ***and***
No active precipitation occurring ***or forecast***
If used—no holdover time allowed
Other allowable methods
Forced air
With or without injected type I fluid
No holdover
Infrared Radiant Heat
Used only in a few places such as Continental hangar at EWR and JFK
No holdover

Two-Step Deicing / Anti-icing
Removes existing accumulation
Prevents (for limited time) recurrence
Deicing fluids allowed as above
Anti-icing
Can ONLY occur in a two-step process
Fluids allowed include
Type I—limited holdover that only applies IF second application used
Type II (non-domestic only) or IV (international or US domestic)
Outside US—may be heated and diluted to provide one-step deice / anti-ice
Must be flushed off if
Holdover has expired
Fluid has failed—contaminated with snow or freezing precipitation
Another coat cannot be applied over failed fluid layer
Must include critical surfaces, to include
Wing leading edges and upper surfaces
Horizontal stabilizer and elevator upper surfaces
Vertical stabilizer and rudder

Stabilizer Requirements
Set for takeoff specified on TPS/closeout
Verify trim is in the green band

Cold Soaked Fuel Frost (CSFF)
Generally occurs with large amounts of cold fuel remaining after landing when that fuel has been in cold temperatures for a long period
Conditions favoring formation:
- Large amount of fuel still in wing tanks after landing
- Frost/ice under wing outboard of measuring stick 4
- Cockpit fuel temperature gauge reads less than 4°C
- Can occur with ambient temperatures as high as 15°C

If CSFF suspected, check upper wing for ice or frost
AOM 15.8.2, p. 15-26 to 15-15-27

Lowest Operational Use Temperature (LOUT)
Defined as actual fluid freeze point plus fuel specific temperature buffer. See AOM 15c.6 & fuel holdover tables

Engine Run-Up Summary

When engine anti ice required and OAT is ≤ 3°C (37°F), ***or***
If there is increased fan vibration due to fan ice accumulation ....do an engine sunup to minimize ice buildup:
Check area behind aircraft is clear
Run up to minimum 70% N1 for 30 seconds at least every 30 minutes
If conditions do not allow runup as above, set a thrust level as high as practical for approximately 30 seconds
Fan blade ice build up is cumulative. If the fan spinner and fan blades were not deiced prior to taxi out, the time the engines were operating during the taxi in should be included in the (30 min. 3AA-3PX) or (60 min. 3RA-3VM) interval.........AOM 15.8.5

Operating in freezing rain, freezing drizzle or freezing fog:
Run up to 70% N1 for approximately 1 second at intervals no greater than once every 10 minutes
Enhances ice shedding......................AOM 15.8.4, p. 15-32

Engine Anti-Ice Use:

Ground—Must be **ON** immediately after each engine is started and remain on during all ground operations when icing conditions exist or are anticipated......................AOM 30.4

In Flight
Use when conditions within icing condition definition—visual moisture plus temperature less than 10°C (50°F) unless in cruise or climb below -40°C (-40°F).
WARNING —do not rely on structural ice or visual cues to begin engine anti-ice use
CAUTION —Do not use engine anti-ice when OAT (ground) or TAT (inflight) is above 10°C (50°F).......AOM 15.8.6; 18.5.2

Wing Anti-Ice Use

Deicing or anti-icing use allowed
- ◆Deicing—allow ice to build up, then use to remove
- ◆Anti-icing—prevention of build up

Ground—Use between engine start and takeoff ***when:***
Icing conditions exist or are anticipated, ***unless***
Aircraft is, or will be protected by application of
- ◆Type II or Type IV fluid
- ◆In compliance with approved ground deicing program

In Flight—Use when ice accumulation is observed
Ice accumulation on the flight deck window frames, windshield center post, or on the windshield wiper arm is an indication of structural icing conditions

WARNING Do not use wing anti-ice instead of ground de/anti-icing

CAUTION Do not use wing anti-ice when OAT (ground) or TAT (inflight) is above 10°C (50°F) ....AOM 15.8.6; 18.5.2
- ◆If used above FL350, may cause bleed trip off and possible loss of cabin pressure..............AOM 15.8.6
- ◆Prolonged operation in icing conditions with leading & trailing edge flaps extended is prohibited...AOM 18.11.1
- ◆Holding in icing conditions with flaps extended is prohibited ....AOM 15.8.6
- ◆ Stall speeds may vary due to ice accumulations on the aircraft, especially on the wings. ....AOM 15.8.6

Taxi Out

Flaps/slats should remain retracted during taxi until reaching takeoff runway if:
- ◆Taxiway is contaminated
- ◆Freezing precipitation continuing after de-icing / anti-icing

Operating flaps during low temperatures
Monitor flap position indicators for positive movement
If flaps stop, immediately place Flap / Slat Handle in the position indicated
If flaps are left retracted during taxi to avoid slush/ice:
- ◆Complete Before Takeoff checklist after flaps are in the takeoff configuration
- ◆Check full travel of all flight controls
- ◆Verify trim moves freely after deicing complete

Nose steering—Cycle in both directions to warm fluid

Holdover Considerations

Loss of effectiveness / fluid failure
Fluid has failed if any ice or snow is seen on treated surfaces
Deicing / anti-icing must be re-accomplished prior to takeoff

Inspections
Failure normally starts on leading or trailing edges of wing
If A/C heading is downwind, mid-chord area tends to show fluid failure first
Leading edges and upper surfaces must be checked for evidence of fluid failure
Evidence of Type II or IV fluid failure
Gray or white appearance and buildup of ice crystals in or on top of the fluid
Progressive surface freezing or snow accumulation
Dulling of surface reflectiveness

Accomplish a Cabin/External Pretakeoff Contamination Check
Whenever any doubt exists that the aircraft is free of adhering contamination after F/D Pre-TO Check, OR
If the holdover time has expired
Flightdeck Pretakeoff Check ....AOM 15b.4.5
Cabin/External Pretakeoff Contamination Check .AOM 15b.4.6

Pre-takeoff contamination check
May be conducted from outside the aircraft by "authorized deicing personnel"
T/O must occur within five minutes of inspection....AOM 15b.4.6

Precipitation for which Takeoff is Prohibited
Heavy ice pellets or Hail
Moderate or heavy freezing rain ....AOM 15a.1.3

Takeoff

Slush or snow
Takeoff performance is degraded
Must refer to slush / snow accumulation adjustments in Performance section

Align aircraft with runway before applying **T/O thrust**

Run-Up Requirement
__If__ taking off in moderate to severe icing conditions ***OR*** when temperature is 3°C or below
__Then__ run engines to 70% and check for stable engine operation ***prior to*** brake release
If airplane starts to slide, begin takeoff roll and crosscheck engine instruments for reasonableness during takeoff

Wing Anti-ice (if required), normally turned on after first thrust reduction on departure ....AOM 15.8.6

Steering effectiveness
Forward pressure on yoke below 80 knots increases nose wheel effectiveness
Smooth rudder pedal input used if deviating from centerline
PM should monitor engines for symmetric thrust
Limit rudder pedal deflection to ½ on "very slippery runways"
Lag in rudder pedal steering and nose wheel skidding should be anticipated on icy runways
Rudder becomes effective between 40 -60 knots

Climb, Enroute, Holding and Approach

Prohibited: Holding with flaps extended in icing conditions

Not Recommended:
Prolonged operation in icing conditions with leading and trailing edge flaps extended
Use of wing anti-ice above approximately FL350
May cause bleed trip off
Possible loss of cabin pressure can result

Wing Anti Ice On (if required) above 1000' AGL

Engine Run Up Requirement
__Ground and before T/O__—Used to minimize ice-buildup when:
In icing conditions and the OAT is at or below 3°C, OR
Increased fan vibration due to fan ice accumulation
__Inflight__—Fan Ice removal
May be required when operating in prolonged moderate to severe icing conditions
Run up engines to a minimum RPM of 80% N1 for approximately 1 second
If engine vibration ≤ 4.0 units after thrust is reduced, repeat at 15 minute intervals
If engine vibration > 4.0 units, see next paragraph
....AOM 15.8.6, p. 15-37

High Engine Vibration
Defined in QRH as over 4.0 vibration reading accompanied by airframe vibration
Can be caused by ice buildup on engine spinner
Engine RPM is cycled from 45% to 80% N1
Refer to QRH "High Engine Vibration".QRH ENGINES, APU-7.17

Fuel Temperature Minimums
Jet A: -37°C
Jet A1: -43°C
Cooling rate
Should not exceed 3°C per hour
In "the most extreme conditions," can reach a maximum change rate of 12°C per hour, however
Possible methods to handle fuel approaching freezing point:
Descend to a warmer air mass
Divert to a warmer air mass
Increase Mach number

Ice Crystal Icing
Threat to engines due to ice ingestion and damage
See definition and characteristics ....AOM 15.3
Conducive condition recognition ....AOM 15.3.1
Prevention—Avoid flying directly over large convective cells with red radar returns, even if no returns at A/C Altitude
....AOM 15.3.2

Landing

Flaps 40 recommended for landing on **slippery runways**
....AOM 15.8.8

Recommended technique
Land on speed
Touchdown at planned point; firm better than a "grease job"
Keep NW firmly on runway with elevator; ***but*** excessive down elevator decreases weight on main gear and reduces braking efficiency
Use as much of the runway for roll-out as needed to slow airplane to a safe taxi speed before turning off a wet / slippery runway
Autobrakes 3 or MAX ....AOM 9.1.1, p. 9-4
When taking over manual braking—Use full brake pedal pressure; do not pump brakes ....AOM 12.1.6, p. 12-26
....AOM 15.7.7, p. 15-20
With crosswinds, land in crab and correct after touchdown
....AOM 12.2.1, p. 12-34
Maintaining directional control is the highest priority
In an emergency, use max reverse thrust, if required, to stop in the remaining runway...........................AOM 15.7.7

Airplane General

Overview

Dimensions

Width (With winglets)........117'-5" (35.79m)
Gear, (main gear center of trucks)..................18'8" (5.70m)

Height—Tail ....41'-2"(12.50m)
Winglets..........................21'-1 (6.42m)

Length—Fuselage ....124'-9" (38.00m)
Total aircraft (to aft tips of elevators) ....129'6" (29.50m)
Nose gear to main gear .....51'1" (15.60m)

Turning Radius
Min pavement width, 180° turn..............79.1 Feet (24.1 Meters)
Wing Tip Radius ....75.2 feet (22.9 meters)
Other Radii—..................see AOM 1.10.2

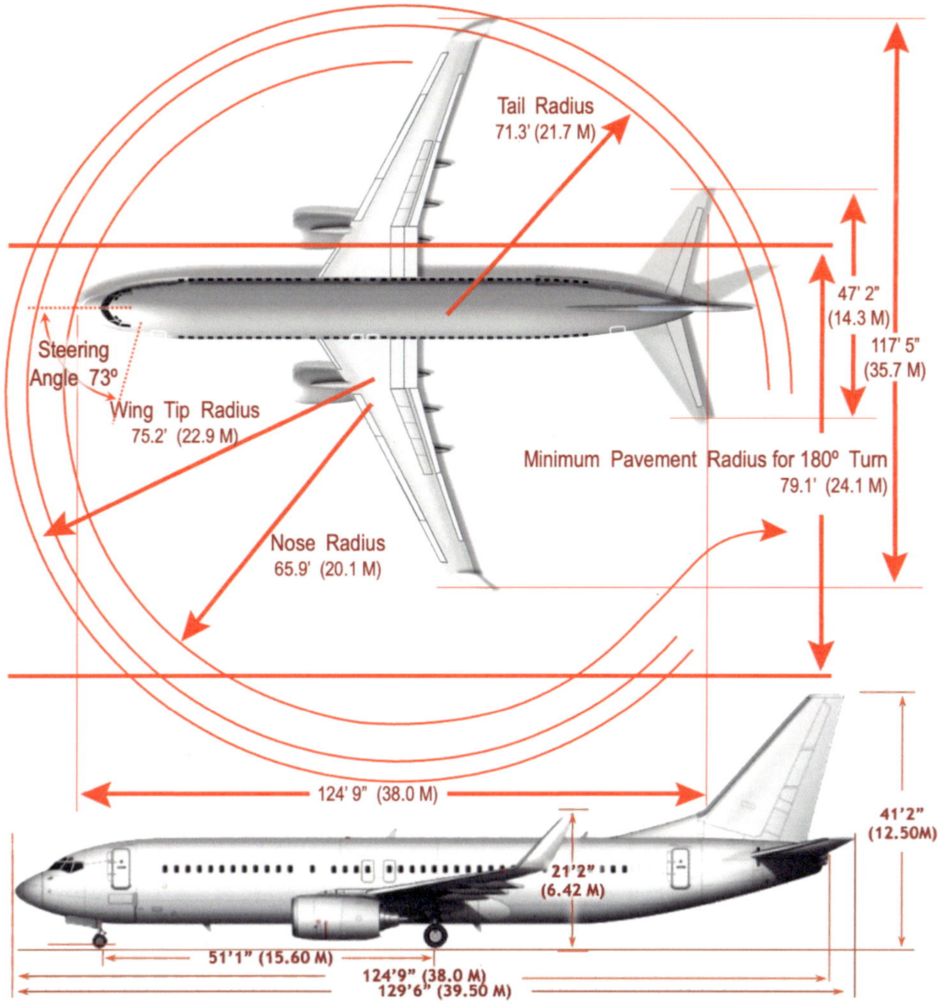

Side silhouette courtesy of ***Norebbo Stock Illustration and Design***, www.norebbo.com. with permission.

Lighting & Status Indications

Emergency (EMER) EXIT LIGHTS Switch

OFF—Prevents Emergency Lights System operation if airplane electrical power fails or is turned off

ARMED—(Guarded) all Emergency Lights illuminate automatically ***if***

- Power to DC bus No. 1 fails ***or***
- AC power is turned off

ON—All Emergency Lights illuminate

NOT ARMED Light—Emergency exit light switch not **ARMED** (either **OFF** *or* **ON**) .... **MC** **OVERHEAD**

Passenger Cabin Switch

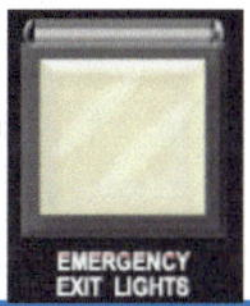

Location—Aft F/A's Panel
OFF—(Guarded)—Cockpit switch controls
ON—All Emergency Lights ON
Bypasses flight deck control

FASTEN BELTS Switch

Cockpit Switch
AUTO—**FASTEN SEAT BELTS** and **RETURN TO SEAT** signs illuminate automatically when gear *OR* flaps are extended
Lights extinguish when gear *AND* flaps are retracted
ON—**FASTEN SEAT BELTS** & **RETURN TO SEAT** signs illuminated

No Electronic Devices Switch

3LC and Subsequent; **AUTO**—signs on with gear down

Exterior Door Annunciations

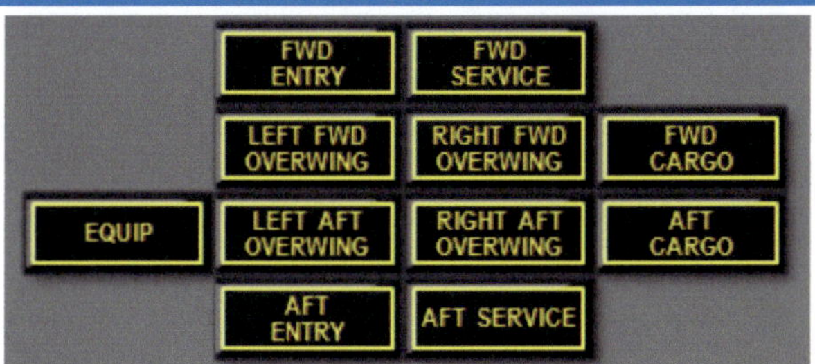

Entry Doors—Related door not closed & locked .... **MC** **DOORS**

Overwing Exit Annunciations—Illuminated if
- Related overwing exit is not closed and locked ***or***
- Related flight lock failed to engage when commanded

(See next page)

Reinforced Flight Deck Door Control Panel

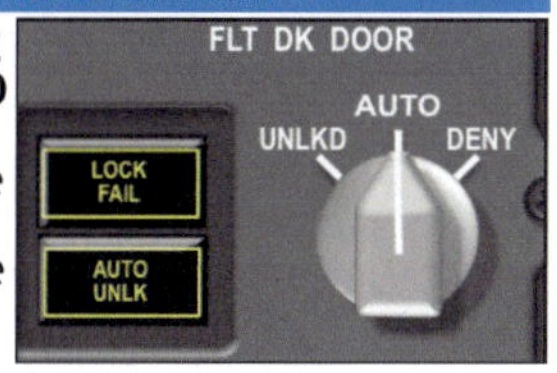

Flight Deck Door LOCK FAIL Light
ON—Flight Deck Door Control in **AUTO** position ***and***
- Door locking solenoid ***not*** in the locked position ***or***
- Power Cutoff Switch on the chime module is **OFF**.

AUTO Unlock (UNLK) Light
Illuminated—Correct access code entered in keypad
On the flight deck:
- **AUTO UNLK** light ON continuously ***and*** chime sounds
- 20 seconds—chime sounds again
- 40 seconds—continuous chime ***and*** flashing **AUTO UNLK** light
- 60 seconds—green LED above cabin keypad illuminates
- Door unlocks for 5 seconds, then re-locks

Flight Deck Door Control—Spring loaded to **AUTO**
UNLKD—Push-to-rotate, momentary position—Unlocks door.
Door remains unlocked while switch is held in **UNLKD** position.
AUTO—With DC Bus 2 power on, door locks whenever closed
Door can be opened from flight deck side using handle
Allows opening of door as above unless **DENY** selected
DENY—
- Rejects keypad entry request
- Cancels alert
- Prevents further access code entry for 5 minutes.

Flight Crew Oxygen System

Maximum pressure may be as high as 1850 psi

Minimum Pressures
dependent on C/P occupants and temperature

Operational Notes
Crew Oxygen Pressure:
If pressure is < 850 psi but above minimum minimum dispatch pressure for conditions, AML INFO-ITEM entry is required ....AOM 3f.1.3, p.3f-8
Minimum for dispatch ....AOM 18.3.5
Crew Oxygen Check—Pressure should not drop by over 100 psi ....AOM 3f.1.3, p.3f-7

Passenger Oxygen System

Activation—Cabin pressure **14,000' *or***
PASS OXYGEN Switch (overhead panel) is **ON**
Cabin Altitude Light (3DM & Subsequent).......... *See page 30*

When activated

PASS OXY ON Light .... MC OVERHEAD
Masks to drop from the stowage compartments
Oxygen generators are activated when any mask in the unit is pulled down
Pulling one mask down
Causes all masks in that unit to come down
100% oxygen flows to all masks
In-line flow indicator (green) is visible in the transparent oxygen hose whenever oxygen is flowing to the mask
Oxygen flows for approx. 12 minutes; *cannot be shut off*
If passenger oxygen is activated and a PSU oxygen mask compartment does not open, masks may be dropped manually
FASTEN SEAT BELT signs (if switch in **AUTO**) illuminate
Lavatory **RETURN TO SEAT** signs do *not* illuminate
O2 Masks available in lavatories when system activates

Other Emergency Systems

PBE Humidity Indicator

Indicates presence of moisture in the clear plastic package containing the PBE
Blue —No moisture in clear plastic package
Pink —Indicates moisture in the clear plastic package

Rafts

Three 56 person rafts (with overload capacity of 84) are installed in the ceiling compartments of overwater equipped airplanes

One additional 10 person raft kit may be installed on 160 seat cabin aircraft (requirement to be deleted by FAA)

Emergency Locator Transmitter (ELT)

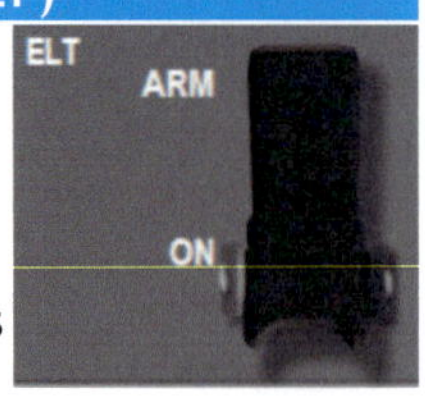

Water activated

Location(s)
Above the left aft overwing exit
First class cabin left aft doghouse (SEL 14)

ELT Panel & Switch (3DM & Subsequent)
ARM—ELT Transmits automatically when its preset G-Load limit is reached
ON—Manually activates ELT

ELT—Light on when ELT <u>activated</u> and <u>is transmitting</u>
Can be reset by cycling switch to **ON** and then **ARM**............................ This information removed from current OMs
737 Cockpit Companion, p. 14 (*Not covered in AOM or ASM*)

Emergency Exit Door Flight Lock System

Power—28 Volt DC

System is designed to ensure that the flight lock will:
◆Automatically lock during takeoff, in-flight, and landing ***and***
◆Unlock on the ground to allow for opening of the door in emergency situations

Automatically Locks
During takeoff, in-flight, and landing when
◆Three of the four entry/service doors are closed ***and***
◆Either engine is running ***and***
◆The airplane air / ground logic indicates that
◆◆The airplane is in the air ***or***
◆◆Both thrust levers are advanced

Automatically unlocks when
Any of the above conditions NOT met ***or***
DC Power is lost

LEFT or **RIGHT OVERWING** Warning Lights .. MC **DOORS**

An emergency exit door is not fully closed and locked *OR*
The flight lock is not engaged, either during the takeoff roll or in-flight

PSEU Light and Flight Lock / Door Warning Lights

Indications .... MC OVERHEAD
◆A flight lock has failed locked *OR*
◆A fault is detected

Light is inhibited:
◆In flight
◆When the thrust levers are advanced toward takeoff power
◆For 30 seconds after landing

When doors are latched and locked, and the flight lock is operating properly none of these lights illuminate:
◆**PSEU** ◆ MC ◆ OVERHEAD

3DM & Subsequent Differences
◆**PSEU** Light and System Annunciator **OVERHEAD** do ***not*** come on for simple faults unless System Annunciator Panel is pressed for recall
◆Resetting **MASTER CAUTION** system extinguishes **PSEU** light
<u>Note</u>: *Additional information is covered in WARNING SYSTEMS in ASM and in this book,* *page 65*

MAX **PSEU Light removed**
Some functions incorporated into **MAINT** light
See additional information in WARNING SYSTEMS in AMS and in this book, *page 65*

Fuel Notes

Ground Use

Minimum Center Tank Fuel Volumes
To turn **ON** center fuel pump on ground ....1000 lbs
To run unattended ....Prohibited—(use wing tank)
Minimum wing tanks fuel if center fuel > 1000 lbs..........FULL
AOM 14f.1.3, p. 14f-4

Minimum Wing Tank Fuel to operate an electric motor-driven hydraulic pump...............1675 lbs
ASM Hydraulics 13.20.2, p. 13.20.5

In Flight Use

Center Tank Fuel Pumps
Climb—if center pump **LOW PRESSURE** light illuminates, affected pump **OFF** (both pumps off if center tank empty)
In level flight, if center tank still contains fuel, center tank pump switches **ON**
When center pump **LOW PRESSURE** light illuminates, affected pump **OFF** (both pumps off if center tank empty)

Wing Tanks
Should remain balanced within ....1,000 lbs
Fuel IMBAL alert
ON at......................................Difference > 1000 lbs
OFF at.....................................Difference < 200 lbs

Air Conditioning and Pressurization

General

Bleed Air Overview

Bleed Air Sources
- ♦Engines ♦APU ♦External air cart / source

Bleed Air Used By
- ♦Air conditioning / pressurization
- ♦Wing and engine thermal anti-icing
- ♦Engine starting
- ♦Hydraulic reservoirs pressurization reservoirs
- ♦Water tank pressurization
- ♦Aspirated TAT probe

External & APU Bleed Sources

External air source—Provides an alternate air source for engine start or air conditioning

APU Bleed Air Valve—permits APU bleed air to flow to the bleed air duct
Valve closes automatically when APU is shut down
DC controlled, pressure operated
Engine 9th stage valve is signaled closed to prevent back pressure on APU bleed valve if
- ♦APU bleed valve open ***and***
- ♦Engine bleed air valves open, ***and***
- ♦Engines operating at idle thrust

DUAL BLEED Light—**ON** when above conditions exist
Engine thrust must be limited to idle to avoid 9th stage opening and resulting backpressure

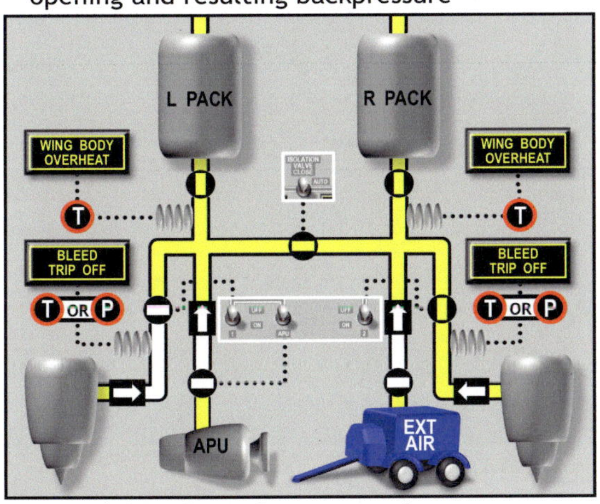

With External Air Cart Connected
Battery should be kept on
Protective circuits are DC
Recommended switch positions:
- ♦Recirculation fan **AUTO**
- ♦Isolation Valve **OPEN**
- ♦APU Bleed air **OFF**
- ♦Packs **AUTO** or **HIGH**

Note: *This information in academic material, **not** in AOM or ASM*

Engine Bleed System Supply

Engine bleed air from the **5th** and **9th stages** of the compressor section
When 5th stage low pressure bleed air is insufficient for system requirements, the high pressure 9th stage bleed opens
High power settings—
Takeoff, climb, and most cruise conditions...
Low pressure bleed air from the 5th stage is adequate; high stage valve remains closed

Engine Bleed Air Valves
Act as pressure regulator and shutoff valves
DC activated, pressure operated
Maintain proper system operating pressure
Reduce outflow if bleed air temperature too high

Bleed Trip Sensors—detect bleed temperature or pressure exceeding predetermined limits
BLEED TRIP OFF Light results

BLEED TRIP OFF—Excessive engine bleed air temperature or pressure ...... MC **AIR COND**

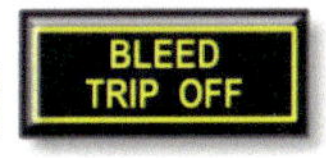

Related engine bleed air valve closes automatically
Requires reset

MAX **BLEED replaces BLEED TRIP OFF**

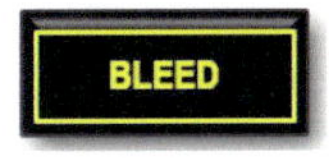

High temperature ***or*** pressure high ***or*** low
Bleed closes automatically ...Requires reset
Bleed air system failure....................May ***not*** be resettable
Incorrect bleed air configuration after takeoff or go-around
Turning either BLEED switch ON extinguishes both BLEED lights if on due to configuration problem

PACK Lights .... MC **AIR COND**

Pack trip off due to overheat* *OR* Failure of both primary and standby pack controllers
**737 Cockpit Companion*, page 89
During Master Caution recall—light **ON** for
Failure of either primary or standby pack control
Extinguishes at Master Caution reset
MAX Light indicates above conditions ***or***:
Pack overheat or dual control failure ***or***
Flow control valve fails in closed position ***or***
Configuration--both pack switches are **OFF** 45 seconds after flap retraction on takeoff

WING-BODY OVERHEAT ..... MC **AIR COND**

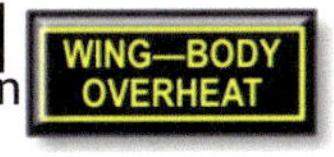

Left light—overheat from bleed air duct leak in any of the following:
- ♦APU Bleed Air
- ♦Left Inboard Wing LE Duct
- ♦Keel Beam
- ♦Left A/C Bay
- ♦Left Engine Strut

Right light—overheat from bleed air duct leak in any of the following:
- ♦Right Inboard Wing LE Duct
- ♦Right A/C Bay
- ♦Right Engine Strut

Duct Pressure Transmitters—Provide bleed air pressure indications to respective (L and R) pointers
Indicator is AC operated
Differences between L and R duct pressure considered normal if cabin pressurization needs are met

Isolation Valve—isolates the left and right sides of the bleed air duct during normal operations
AC powered
Valve is closed when
Isolation Valve Switch in **AUTO** ***and***
Both Engine Bleed Air Switches **ON** ***and***
Both Air Conditioning Pack Switches **AUTO** or **HIGH**
Valve opens when
Either Engine Bleed Air Switch **OFF** ***or***
Air Conditioning Pack Switch **OFF**
Valve position not affected by the APU Bleed Air Switch

Bleed Air Controls

Engine BLEED Air Switches

OFF— Closes engine bleed air valve
ON—Opens engine bleed air valve if engines running

APU BLEED Air Switch

OFF—Closes APU bleed air valve
ON—Opens APU bleed air valve when APU is operating

DUAL BLEED—Two bleed sources are potentially pressurizing the APU bleed supply line ...... MC **AIR COND**

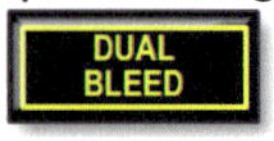

APU bleed air valve open *and either*
Engine No. 1 bleed air switch **ON**, ***or***
Engine No. 2 bleed air switch **ON** with isolation valve open.

ISOLATION VALVE Switch

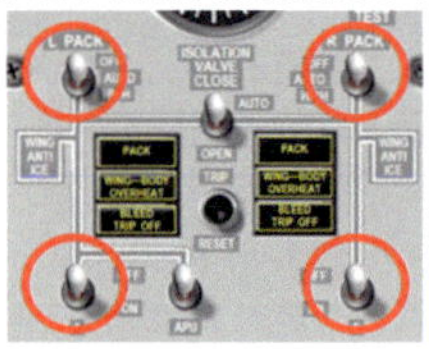

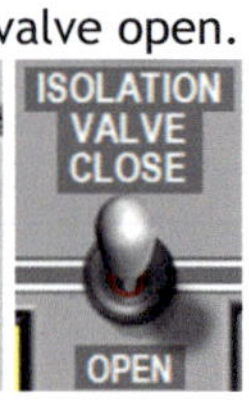

CLOSE—Closes isolation valve
AUTO—Associate with four "corner" switches
Closes isolation valve ***if***:
Both engine Bleed Air Switches are **ON** ***and***
Both **PACK** Switches **AUTO** or **HIGH**
Opens isolation valve ***if***
Either engine bleed air or air conditioning **PACK** switch positioned **OFF**
OPEN—Opens isolation valve

Wing-Body Overheat (OVHT) TEST Switch

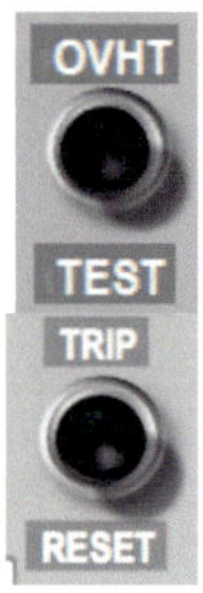

♦Tests wing-body overheat detector circuits
♦Illuminates both **WING-BODY OVERHEAT** Lights

TRIP RESET—When pressed, if fault is corrected
♦Resets **BLEED TRIP OFF**, **PACK** and **ZONE TEMP** lights
♦Related engine bleed air valves open, or related pack valves open, or related air mix valves open
♦Lights remain illuminated until reset

MAX Digital Bleed Air Supply Control

Electronic control, pneumatic operation
Automatic fault detection and isolation
Both packs can be operated from a single bleed air source
See description – ASM 2.20.2 3RA - 3VM Digital Controllers

Air Conditioning System

Introduction

External air source flows into mix manifold
Flight deck air comes directly from left pack upstream of mix manifold
Excess left pack air combines with recirculated air in mix manifold
Mixed air is distributed to cabin sidewall riser outlets

Air Conditioning Packs

Left & Right Packs normally use air from respective engine bleeds
Single pack
Capable of maintaining pressurization and temperatures up to the max A/C certified ceiling
MEL limits single pack ops (if dispatched single pack) to 25,000'MSL...MEL 21-1

Supply Capability

APU—Two packs on the **ground,** or one pack **in flight**
Most external air carts can supply adequate bleed air for two pack operation
Engines should not be used for more than one pack from one engine at any time

Airflow Control

Normal flow—both Pack Switches in **AUTO** and both packs operating
Single pack—automatically switches to "high air flow" to maintain the necessary ventilation rate
Automatic high flow occurs regardless of flap position, air/ground status or number of packs operating ***if***
♦One pack not operating (other goes to **HIGH**) ***or***
♦Both Engine Bleed Air Switches are **OFF** ***and***
♦APU Bleed Air Switch is **ON**
Automatic switching is inhibited when
♦The airplane is on the ground, ***or***
♦Inflight with the flaps extended
Ensures adequate thrust for single engine operation
With A/C Pack Switch in **HIGH**, pack always high flow
APU high air flow rate is available when
♦On the ground ***and***
♦APU Bleed Air Switch is **ON** ***and***
♦Either Pack Switch (or both) in **HIGH**
Provides maximum airflow when APU is the only bleed air source

Air Conditioning Controls and Indicators

ZONE TEMP Lights .... MC **AIR COND**

♦CONT CAB **ZONE TEMP** light indicates duct temperature overheat ***or***
Failure of the flight deck primary ***and*** standby temperature control
♦FWD CAB or AFT CAB—duct temperature overheat ***only***
During Master Caution Light recall:
♦CONT CAB **ZONE TEMP** indicates failure of the flight deck primary ***or*** standby temperature control
♦Either FWD or AFT CAB **ZONE TEMP** indicates failure of the associated zone temperature control
♦Lights will extinguish when Master Caution is reset

Temperature Selector

AUTO—Automatic temperature control for the associated zones
Rotating the control toward C (cool) or W (warm) manually sets the desired temperature
Temperature range is 65˚F (C) to 85˚F (W)
OFF—Closes associated trim air modulating valve
Any SINGLE selector **OFF**—that selector is ignored
All three selectors **OFF**
Selects 75˚F (Left Pack Output) ***and***
65˚F (Right Pack Output)

RAM DOOR FULL OPEN Lights—Indicates ram door in full open position

MAX Lights removed

Air Conditioning PACK Switches

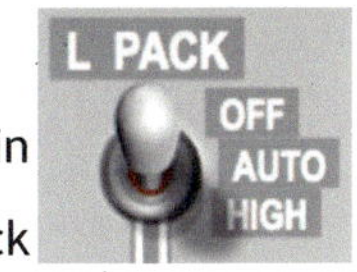

OFF—Pack commanded off
AUTO—With both packs operating, each runs in low flow
- ♦With one pack operating, operating pack in high flow in flight with flaps up
- ♦When operating one pack from APU (both engine BLEED Air Switches **OFF**), runs in high flow

HIGH—Pack regulates to high flow;
Provides maximum flow rate on ground with APU BLEED Air **ON**

TRIM AIR Switch—Commands trim air pressure regulating & shutoff valve open or closed

Recirculation (RECIRC) FAN Switches

OFF—Fan commanded off
AUTO—In-flight
- Left **RECIRC** fan operates if both packs are operating unless either PACK Switch in **HIGH**
- Right RECIRC fan operates if both packs are operating unless both PACK Switches in **HIGH**

On the ground
- The left **RECIRC** fan operates unless both **PACK** switches are in **HIGH**
- The right **RECIRC** fan operates even if both **PACK** Switches are in **HIGH**

Conditioned ground air
Pumps air into mix manifold
Preconditioned air is then distributed throughout the airplane

Cooling Cycle
Starts with bleed air cooled through a heat exchanger
Air then flows to air cycle machine for refrigeration
Processed cold air combines with hot air which has bypassed the air cycle machine, then through
High pressure water separator which removes moisture
Conditioned air then flows into the mix manifold and distribution system

Overheat protection
Provided by temperature sensors in the cooling cycle
Overheat condition causes
- Pack valve to close ***and***
- **PACK** Light to illuminate (*See page 26*)

Pack Temperature Control
2 Electronic controllers—Primary & Standby
If primary pack control fails affected pack is controlled by standby pack control in opposite controller

Failure Indications
- ♦Primary or standby pack control failure
  - ♦♦During Recall .... MC | AIR COND
- ♦Both primary & standby pack controls fail for the same pack
  - ♦♦*Without* Recall .... MC | AIR COND
  - ♦♦Pack keeps operating without control unless excessive temperatures cause it to trip off
- ♦Pack overheats, causing trip off ***AND*** associated PACK light .... MC | AIR COND

Zone Temperature Control

Concept behind Temperature Control

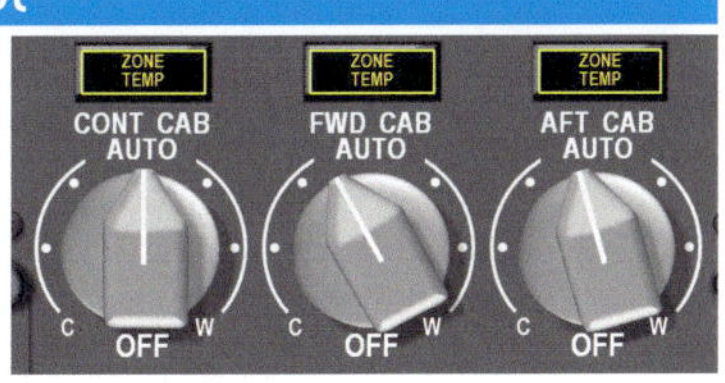

Three zones:
- Flight deck (CONT CABIN)
- Forward and aft cabins (FWD and AFT CABIN)

Two Packs—The *challenge* is controlling three zones with two packs and trim air
Two Electronic Controllers manage this process
- Left electronic controller controls the aft cabin zone ***and*** Provides backup control for the flight deck
- Right electronic controller controls the forward cabin zone ***and*** provides primary control for the flight deck

Desired zone temperature set by individual Temperature Selectors
Selector range is approximately 65°F to 85°F

When all three zones and **both packs** are operating normally,
Pack temperature output for both packs satisfies zone requesting the coolest temperature air
Trim air is blended with cool air to provide requested supply air to each zone
- Respective zone temperature is controlled by adding trim air to the zone supply ducts
- Quantity of trim air is regulated by individual trim air modulating valves

Single pack operation
With **TRIM AIR ON**
- Temperature control is same as two pack operation
- Single operating pack temperature is determined by zone requesting the coolest air

With **TRIM AIR OFF** — Pack tries to produce temperature to satisfy average temperature demanded by all three zones

Zone supply duct air overheat results .. MC | AIR COND
- ♦Associated amber ZONE TEMP Light *and*
- ♦Associated trim air modulating valve closes

Trim air modulating valve may be reopened after the duct has cooled by pressing the TRIP RESET Switch

Zone Temperature Control Modes
Unbalanced Control
Indicates packs are being individually controlled, not in unison
Caused by—any failure affecting supply of trim air
Loss of Flight Deck trim air results
- Left pack supplies conditioned air to F/D at requested temperature
- Right pack output determined by the cabin zone requesting the coolest air

Loss of either (or both) cabin trim air results
- F/D temperature request satisfied by output of the L Pack
- Forward and aft zone temperature demands averaged for control of right pack

Standby Pack Average Temperature
Occurs when zone controls and primary pack controls fail
Standby pack controls command packs to produce air temperatures that will will satisfy the average temperature demand of the two cabin zones.
The trim air modulating valves will close.
Flightdeck zone temperature selector has no effect

3RA-3VM Dual IASC Failure
Occurs with dual failure of the IASCs resulting in both primary and backup controls unavailable:
Adjusting Zone Temperature selectors will not change temperature
Pack Temperature Control valves and the three Zone Trim Air valves remain in the last commanded positions
Pack overheat protections are functional and automatically shut down Air Conditioning Pack if a pack overheat trip occurs
If control cabin (flightdeck) or passenger cabin becomes excessively hot or cold, the flight crew should perform the Cabin Temp Hot or Cabin Temp Cold non-normal checklist.

Fixed Cabin Temperature
3AA-3PX If all Temperature Selectors are positioned OFF, the pack controls will cause the
- Left pack to maintain a fixed temperature of 75°F (24°C)
- Right pack to maintain 65°F (18°C) as measured at the pack temperature sensor.

3RA-3VM If all Temperature Selectors are OFF, IASCs direct
- Left pack to maintain a fixed temperature of 75°F (24°C)
- Right pack to maintain 65°F (18°C) (measured at the pack temperature sensor)

Pack Controller Failure Effects
Failure of primary flight deck temperature control
Note: *This function is handled by the RIGHT electronic controller*
Automatic switch to back up control
ZONE TEMP Light (**CONT CAB**) comes on and remains on after Master Caution Recall
Failure of *both* the primary *and* standby controls—**ZONE TEMP** light comes on *without* recall automatically

Failure of the FWD *or* AFT cabin temp control
Associated trim air modulating valve closes
Temperature Selectors operate normally, *but*
Temperature Selector settings of the two passenger cabin zones are averaged
Indication—Respective **ZONE TEMP** light illuminates with Master Caution Recall

Standby Pack Average Temperature

If all zone controls *and* primary pack controls fail
- Standby pack controls command packs to produce air temperatures to satisfy average temperature demand of the two cabin zones
- Trim air modulating valves close
- Flight deck zone Temperature Selector is deactivated

Air Conditioning Distribution

General
- Conditioned air is collected in the mix manifold
- Air temperature is directly related to temperature selector setting
- Overheat detection provided by temperature sensors located downstream of packs and mix manifold
- Overheat condition causes closing of appropriate trim air modulating valve *and* ZONE TEMP light on

Flight Deck
- All air is from L Pack
- F/D only uses small portion of L Pack air
- Rest goes to mix manifold
- Outlets allow air as desired

Passenger Cabin
- Air supply distribution system consists of
  - Mix manifold
  - Sidewall risers
  - Overhead distribution duct
- Sidewall risers feed overhead distribution duct
- Overhead distribution duct
  - Extends from the forward to aft end of ceiling along airplane centerline
  - Supplies the sidewall diffusers

Recirculation Fan
- Reduces pack load and engine bleed air demand
- Air sources—passenger cabin and electrical equipment bay
- Draws air to the forward cargo bay
- Air is filtered and recirculated to the mix manifold
- Fan motors are AC
- Fans operate only if respective switch is **AUTO**
- In flight
  - L Fan operates if both packs are operating unless *either* PACK Switch is in **HIGH**
  - R Fan operates if both packs are operating unless *both* PACK Switches are in **HIGH**
- On the ground
  - L fan operates unless *both PACK Switches* are in **HIGH** ***and***
  - R Fan operates even if both PACK Switches are in **HIGH**

Equipment Cooling

Overview

Cools electronic equipment in
- ♦Flight deck
- ♦E & E bay

Consists of a **supply duct** and an **exhaust duct**
- Each duct has a normal fan and an alternate fan

Supply duct—Supplies cool air to the flight deck displays and electronic equipment in the E & E bay

The exhaust duct—
- Collects and discards warm air from
  - ♦Flight deck displays,
  - ♦Overhead and aft electronic panels,
  - ♦Circuit breaker panels in the flight deck,
  - ♦Electronic equipment in the E & E bay

Equipment Cooling Control

Equipment Cooling OFF Light
- Indicates loss of airflow due to related equipment cooling fan failure
- Selection of alternate fan
  - Should restore airflow ***and***
  - Extinguish the **OFF** Light within 5 seconds

3DM and subsequent
- With forward cargo fire warning—
  - Equipment cooling exhaust fan is shut off ***and***
  - The equipment cooling exhaust **OFF** light is inhibited
- Overtemperature on the ground—nose wheel well crew call horn sounds

Inoperative Equipment
- See QRH 2.25-2.27 for lists of equipment inoperative after 30, 60, 90 &120 minutes.

Forward Cargo Compartment
- Recirc fans circulate air from passenger cabin around lining of forward cargo compartment
- When overboard exhaust valve is closed,
  - Additional inflight heating provided
    - Exhaust air from the equipment cooling system
    - Added to recirc fan air and diffused into lining of the forward cargo compartment

Equipment Cooling Panel

Equipment (EQUIP) COOLING SUPPLY and **EXHAUST Switches**
- **NORM**—Normal cooling supply/ exhaust fan activated
- **ALTN**—Alternate cooling supply/exhaust fan activated

Equipment Cooling OFF Light—No airflow from selected respective normal or alternate fan......... **MC** **OVERHEAD**

3RA-3VM **EQUIP SMOKE Light**
- smoke detected in equipment cooling system
- Added to bottom of panel depicted above

Pressurization

Overview

Cabin pressure control system
- Two identical automatic controllers
- Active when switch in **AUTO** or **ALTN**
- Backup—manual (MAN) pilot-controlled mode
- Air Supply—bleed air from air conditioning system
- Pressurization and ventilation controlled by modulating
  - Outflow valve ***and***
  - Overboard exhaust valve

Pressure Relief Valves
- Safety relief enabled through two pressure relief valves
  - Limiting differential pressure is 9.1 psi
- Negative relief valve—prevents external pressure from exceeding internal pressure

Cabin Pressure Controller

Rate-controls pressure
- in climb to 8,000' cabin altitude at 41,000' airplane maximum certified ceiling

The cabin pressure controller modes:
- **AUTO**—normal mode of operation
- **ALTN**—Automatic pressurization control; the alternate mode of operation
- **MAN**—Direct, manual system control
- All three modes use DC power

Input data to controllers
- ADIRUs provides Air data including
  - ♦Ambient static pressure
  - ♦Baro corrected altitude
  - ♦Non corrected altitude
  - ♦Calibrated airspeed
- Stall management computers provide thrust lever position
- Air/ground sensors provide appropriate data

Pressurization Outflow

Outflow Valve—main overboard exhaust exit for air circulated through the passenger cabin

Small amounts of air also exhausted through
- Toilet and galley vents
- Miscellaneous fixed vents
- Seal leakage

Overboard Exhaust Valve
- Open on ground and in flight with
  - Low differential pressure
  - Allows warm E&E bay air to discharge
- Closed in flight with
  - Higher cabin differential pressures
  - Causes diffusion of E&E bay air to lining of forward cargo compartment
- Driven open if either Pack Switch is **HIGH** ***and***
  - Right recirculation fan is **OFF**
  - Allows for increased ventilation in the smoke removal configuration
- Smoke Mode commanded (3DM and subsequent) during pressurized flight with Forward Cargo Fire Alarm
  - Opens Exhaust Valve when either pack is in **HIGH** and the Recirc Fan is **OFF**
  - Allows increased ventilation for smoke removal

........................***737 Cockpit Companion***, page 99

Auto Mode Operation
- Two identical controllers
  - Alternately sequenced as primary operational controller for each new flight
  - Other automatic controller is immediately available backup
- Altitude inputs
  - ♦**FLT ALT** (flight or cruise altitude)
  - ♦**LAND ALT** (landing or destination airport altitude)
  - Takeoff Altitude is read continuously when on the ground
- Air/ground safety sensor signals whether airplane is on the ground or in the air
- On the ground
  - At lower power settings—cabin is depressurized by driving outflow valve full open
  - At higher power settings—cabin begins to pressurize
- **OFF SCHED DESCENT** Light ..... MC AIR COND

  - On if airplane begins to descend before reaching preset cruise altitude
  - Controller programs cabin to pressurize to takeoff field elevation without further pilot inputs
  - Automatic abort feature is deactivated If **FLT ALT** indicator is changed
  - To extinguish—reset **FLT ALT** window to actual altitude (or lower) ....QRH AIR SYSTEMS 2.8
- **AUTO FAIL** Light on for of the following ...... MC AIR COND

  - ♦Loss of DC power
  - ♦Controller fault
  - ♦Outflow valve control fault
  - ♦Excessive differential pressure (> 8.75 psi)*
  - ♦Excessive rate of cabin pressure change (± 2000 sea level feet / minute)*
  - ♦High cabin altitude (above 15,800 feet) if controller is not responding properly
  - *If controller is not responding properly
  - With light on, pressure control automatically transfers to the other auto controller (**ALTN** mode)
  - Moving the selector to **ALTN** position
    - Extinguishes **AUTO FAIL** Light
    - **ALTN** Light remains on to indicate single channel operation

Manual Mode Operation
- Manual control of the cabin altitude is used if both the **AUTO** and **ALTN** modes are inoperative
- Outflow Valve Position Switch is used to modulate the outflow valve
- Separate DC standby system-powered DC motor drives the outflow valve at a slower rate than the automatic modes
- Outflow valve full range of motion takes up to 20 seconds
- **MANUAL** Light on with mode selector in the **MAN** position

Alternate (ALTN) Light—Pressurization system is in alternate automatic mode

- Pressurization mode selector in **ALTN** ***or***
- Both **ALTN** and **AUTO FAIL** Lights on—single controller failure and automatic transfer to **ALTN** mode

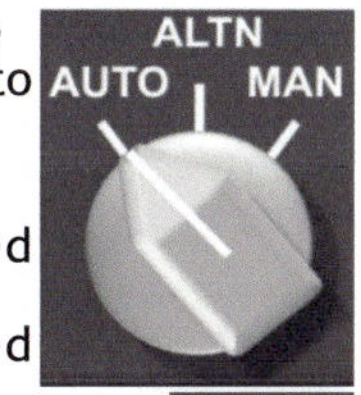

Pressurization Mode Selector
- **AUTO**—Pressurization system controlled automatically
- **ALTN**—Pressurization system controlled automatically using **ALTN** controller.
- **MAN**—Pressurization system controlled manually by Outflow Valve Switch
  - Both auto controllers bypassed

Altitude (ALT) HORN CUTOUT Switch
- Silences intermittent cabin altitude warning horn
- Horn sounds when cabin reaches 10,000' MSL

Cabin Altitude Warning Light

- On if cabin altitude exceeds 10,000'
- Resets to OFF when cabin altitude pressurizes below 10,000'
- Activation simultaneous with aural warning intermittent horn for **CABIN ALTITUDE** alert

Ram Air System

Provides cooling air for the heat exchangers
- Controlled by packs using ram air inlet doors
- Doors move to full open position for max cooling
  - ♦On the ground, ***or***
  - ♦During slow flight with the flaps not fully retracted
- In cruise, the doors modulate between open and closed
- RAM DOOR FULL OPEN light indicates a ram door is fully open

Deflector doors forward of the ram air inlet doors
- Prevent slush ingestion prior to liftoff and after touchdown
- Extend when air-ground safety sensor senses A/C on ground

Recirculation Fans

Fans Operate when their respective switch is in **AUTO** with the following exceptions:

	Left	Right
In Flight		
Both Packs Operating in Auto	ON	ON
One PACK in **HIGH**	OFF	ON
Both PACKs in **HIGH**	OFF	OFF
Ground		
Both Packs Operating in Auto	ON	ON
One PACK in **HIGH**	ON	ON
Both PACKs in **HIGH**	OFF	ON

Pressurization Key Cabin Altitudes

Cabin Altitude Warning Horn ....> 10,000'
Automatic Mask Deployment ....14,000'
Pressurization Auto Fail Triggered ....15,800'

Anti-Ice and Rain Protection

Overview

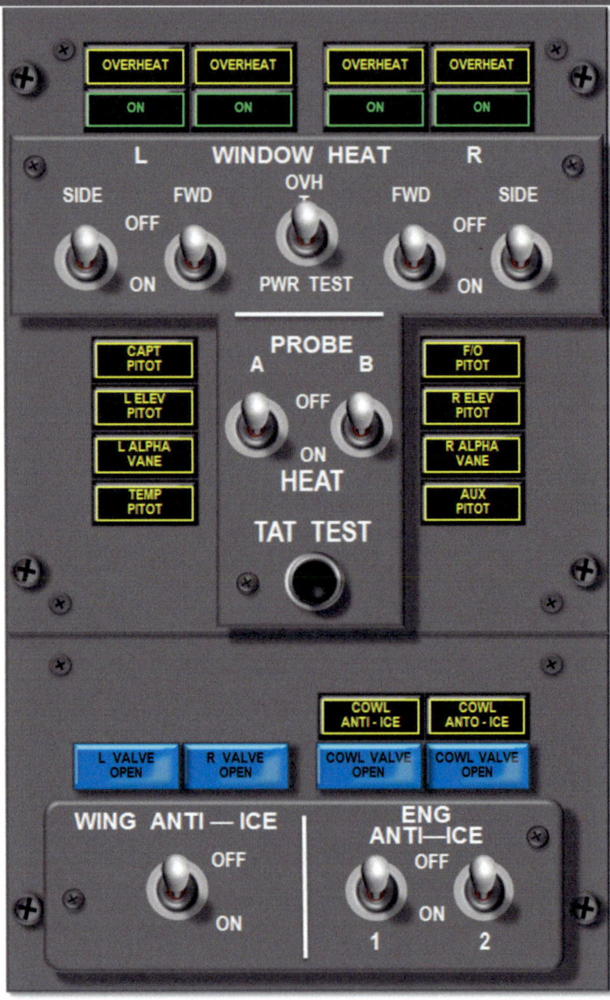

Systems

Ice & Rain Protection

- ♦Thermal anti-icing (TAI)
- ♦Electrical anti-icing
- ♦Physical removal & repellent

Sub-Systems

- ♦Flight Deck Window Heat
- ♦Windshield Wipers
- ♦Probe and Sensor Heat
- ♦Engine Anti-Ice System
- ♦Wing Anti-Ice System
- ♦Permanent Rain Repellent

Window & Probe Heat

Flight Deck Window Heat

F/D Windows heated: 1, 2, 4 and 5 (if installed)
Windows 4 and 5 (L and R) removed from all AA aircraft

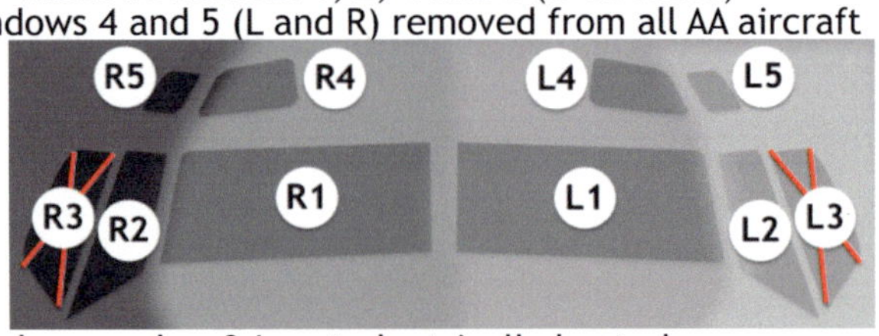

Window number 3 is not electrically heated

Operation

FWD WINDOW HEAT—control heat to L1 & R1
SIDE WINDOW HEAT—control heat to L & R 2, 4, & 5 (if installed)
Temperature controllers maintain window numbers 1 and 2 at correct temperature for maximum strength in event of bird impact

Overheat Control

Window 1 & 2 power is automatically removed
Window 4 (& 5 if installed)—thermal switch opens and closes to maintain temperature

Windshield Wipers

Wipers & permanent rain repellent coating repel water
Windshield scratching occurs if wipers are operated on a dry windshield

Probe and Sensor Heat

<u>Heated</u> (unless on standby power)
- Pitot probes
- Total air temperature probe
- Alpha vanes are electrically heated

<u>Not heated</u>—Static ports

Engine Anti-Ice System

Engine bleed air thermal anti-icing
Prevents ice formation on engine cowl lip
May be operated on the ground and in flight

Engine Anti-Ice System Operation

Each cowl anti-ice valve
- Electrically controlled
- Pressure actuated

Positioning the **ENG ANTI-ICE** Switches to **ON**:
- ♦Routes engine bleed air to cowl anti-ice valve for cowl lip anti-icing
- ♦Sets stick shaker logic for icing conditions
  - Minimum maneuver speed bars adjusted
  - FMC displayed VREF ***not*** adjusted automatically
  - Logic and indications return to normal when engine anti-ice is turned **OFF** ***if*** wing anti-ice has not been used in flight
- ♦If cowl anti-ice valve fails to move to commanded position
  - **COWL VALVE OPEN** Light remains bright blue
  - Amber TAI displays on CDS after short delay

COWL ANTI-ICE Light—excessive pressure in duct from cowl anti-ice valve to the cowl lip........................ MC **ANTI-ICE**

Controls

Window Heat ON Lights

Illuminated (green) - Window heat is being applied to selected window(s)
Extinguished -
- ♦Switch is **OFF**, ***or***
- ♦An overheat is detected, ***or***
- ♦***A system failure has occurred***

Window **OVERHEAT** Lights .... MC **ANTI-ICE**
On—Overheat condition is detected
<u>Note:</u>—**OVERHEAT** *Lights also illuminate if electrical power to window(s) is interrupted*

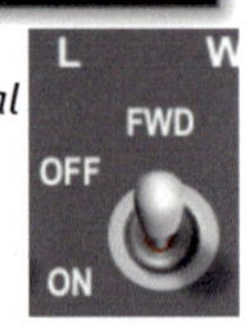

WINDOW HEAT Switches

ON - Window heat is applied to selected window(s)
OFF - Window heat not in use

WINDOW HEAT TEST Switch (spring-loaded neutral)

OVHT - Simulates an overheat condition
PWR TEST - Provides a confidence test

Probe Heat

Probe Heat Lights .... MC ANTI-ICE

On—related probe not heated
On standby power, these Lights ***do not*** indicate system status

Probe Heat Switches

ON—Powers to heat-related system
OFF—Power off

TAT Test Switch

Press—Electrically powers **TEMP PROBE** on ground

Windshield Wiper Switches

Four Settings

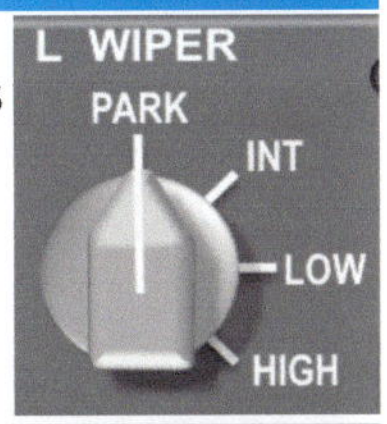

PARK—Turns off wiper motors and stows wiper blades.
INT—Seven second intermittent operation.
LOW—Low speed operation.
HIGH—High speed operation.

Engine & Wing Anti Ice

Engine Anti Ice

COWL ANTI-ICE Lights .... MC ANTI-ICE

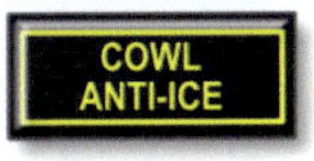

On—Overpressure in duct downstream of engine cowl anti-ice valve (Also covered above)

COWL VALVE OPEN Lights

<u>Bright</u>—Related valve in transit ***or*** valve position disagrees with related **ENGINE ANTI-ICE** switch position
<u>Dim</u>—related valve is open (switch **ON**)
<u>Dark</u>—related cowl anti-ice valve is closed (switch **OFF**)

Engine Anti-Ice Switches

ON—Related engine anti-ice valve is open
Stick shaker logic set for icing conditions
OFF—Related engine anti-ice valve is closed
Stick shaker logic returns to normal if wing anti-ice has not been used in flight.

Thermal Anti-Ice Indications

Illuminated -
Green TAI—Cowl anti-ice valve(s) open
Amber TAI—Cowl anti-ice valve not in position commanded by related Engine Anti-Ice Switch

MAX Core Anti Ice

EECs control system
No flight crew input or indication during normal operations
ENG ANTI-ICE Light—<u>cowl</u> thermal anti-ice system is inhibited due to:
System failure ***or***
An engine <u>core</u> anti-ice valve fails closed

MAX COWL ANTI-ICE Light

Overpressure in duct downstream of engine cowl anti-ice valve

MAX COWL VALVE Light

On (momentary)—related cowl anti-ice valve is in transit
On (steady)—cowl anti-ice valve position disagrees with related ENGINE ANTI-ICE switch position
Off—related cowl anti-ice valve is
Closed (switch **OFF**), ***or***
Open (switch **ON**)

Wing Anti-Ice System

Deices three inboard LE slats

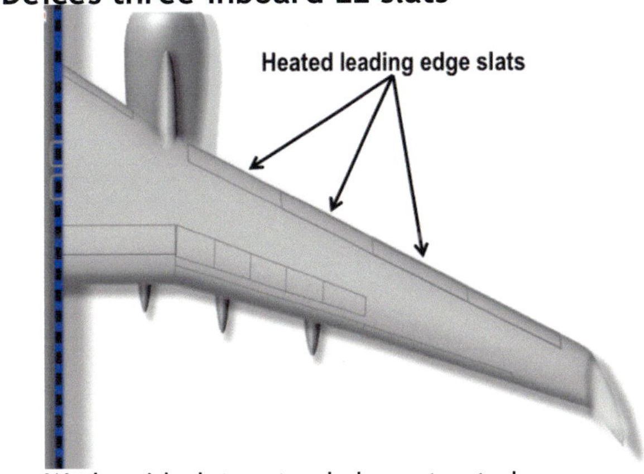

Works with slats extended or retracted

Does not protect

Leading edge flaps
Outboard leading edge slats
Winglets

Wing Anti Ice Switch & System Operation

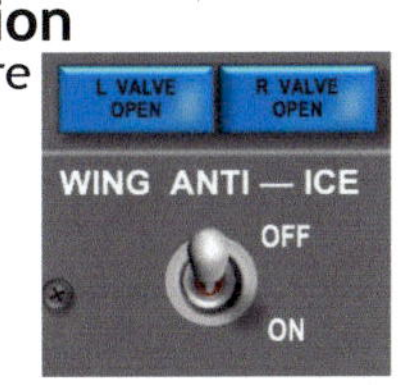

OFF - Wing anti-ice control valves are closed.
<u>Ground</u>—**WING ANTI-ICE** Switch **ON**
Opens both control valves ***if***
<u>Thrust</u> on both engines is below takeoff warning activation setting
and
<u>Temperature</u> in both wing distribution ducts is below thermal switch activation temperature
Both valves close if
thrust on either engine is above takeoff warning setting ***or***
temperature in either duct is sensed as duct over temperature
Valves automatically continue to open and close as conditions above dictate
Switch remains **ON** while on the ground
At liftoff, switch trips **OFF**
<u>In flight</u>—**WING ANTI-ICE** Switch **ON**
Opens both control valves
Sets stall warning logic for icing conditions
Stall warning logic adjusts minimum maneuver speed bars on airspeed indicators (stick shaker)
FMC VREF is **<u>not</u>** adjusted automatically
Logic remains set for the remainder of the flight, regardless of changes of **WING ANTI-ICE** Switch position
Ground use duct temperature and thrust setting logic are disabled in flight and have no effect on control valve operation in flight

Wing Anti-Ice VALVE OPEN Lights

Bright—Related wing anti-ice control valve in transit, or,
Anti-ice control valve position disagrees with **WING ANTI-ICE** Switch position
Dim—Related wing anti-ice control valve is open (switch **ON**)
Extinguished—Related wing anti-ice control valve is closed (switch **OFF**)

MAX L and R VALVE Lights

ON if valve disagrees with commanded position
If on for over 6 seconds. .... MC ANTI-ICE
Replaces blue L and R VALVE OPEN lights on -800 Wing Anti-Ice Panel

Autoflight

Editor's Note: *No attempt is made here to describe the autopilot system in detail. That would be beyond the scope of this document. Only key highlights which are easily missed or forgotten will be covered here for review. It is assumed that most pilots interact with this system sufficiently on a day-to-day basis to be familiar with its operation. In addition, the FGS Trainer at the Flight Academy does an excellent job of introducing this system.*

(Newer Airplanes)

General

Automatic Flight System (AFS) Components

Autopilot Flight Director System (AFDS)

Dual system with two flight control computers (FCC)
- Identified as A and B
- For Autopilot use, they send control commands to respective pitch and roll hydraulic servos
- Servos operate flight controls through two separate hydraulic systems

Controlled through Mode Control Panel (MCP)

Only one Autopilot may be engaged at a time
- Possible to use both Autopilots simultaneously on approach, *but* such use is prohibited by limitation

Autothrottle (A/T)

Flight Management Computer (FMC)

N1 Limit computations provided to
- Autothrottles
- Command airspeeds for A/T and AFDS

Optimized lateral and vertical guidance through climb, cruise and descent

Flight Mode Annunciator (FMA) Display

Mode status displayed on each pilot's PFD

Includes active (engaged) and armed indications

Flight Mode Selection Buttons

If illuminated—indicate that the mode can be *deselected*

In some situations, button blanks--Does NOT mean mode is deselected—check FMA to verify engaged mode

Pictures at right depict older (above) and post-A/C 3DM switch configurations

Autopilot (A/P)

Engagement

May be engaged if:

No force is being applied to control wheel

STAB TRIM AUTO PILOT CUTOUT switch is at **NORMAL**

Accomplished by pressing one of the **CMD** or **CWS buttons**

Normally, selected according to the AFDS Master currently engaged

Dual A/P engagement

Possible on approach

Provides flare and touchdown control

Prohibited by limitation on AA airplanes

Single A/P engagement

AA mode

Does not provide touchdown or "go around capability"

When TO/GA is engaged for go around:
- LNAV guidance is available if
  - ◆An LNAV path is available (G/A is programmed in FMC)
  - ◆Airplane is above 50 feet AGL
- Below 50 feet, mode will be track hold
- Engagement of LNAV go-around guidance is possible if **LNAV** shows armed on FMA

STAB OUT OF TRIM Light

3RA-3VM Operates only with autopilot engaged (extinguished with autopilot not engaged)
- On—A/P is not trimming stabilizer properly
  - May not be able to keep up with speed or configuration change

3AA-3PX ON
- In-flight
  - Autopilot is not properly trimming the stabilizer
  - Partial failure of a Flight Control Computer
  - Illuminates only with autopilot engaged.
  - Remains extinguished when autopilot is not engaged.
- On the ground
  - Partial failure of a Flight Control Computer
  - Illuminates after landing when groundspeed < 30 knots.

Disengagement

◆A/P Disengage Switch—always

◆Go-Around (list tailored for AA mandated single approach mode operations)
- Older airplanes—3AA through 3CT: Pressing **TO/GA** Switch with single A/P engaged in **CWS** or **CMD** below 2000 ft
- Newer airplanes—3CU and subsequent: Pressing the **TO/GA** Switch if any of the following:
  - ◆Below 2000 ft ◆ Flaps are ***not*** up ◆GS is engaged
  - ◆**A/P ENGAGE** Switch—Pressing illuminated switch

◆**DISENGAGE** Bar—pressing down (always)

◆Control Wheel Trim Switch—Activating either side

◆**STAB TRIM AUTOPILOT CUTOUT** Switch—Moved to **CUTOUT**

◆IRS system—(L or R) failure or **FAULT** Light illuminated

◆Electrical—Loss of electrical power ***or*** a sensor input which prevents proper operation of engaged A/P and mode ***or***
- Power interruption to AFDS may cause disengagement
- Reengagement possible after power restored

◆Hydraulics—loss of respective hydraulic system pressure
- *Loss of system A engine-driven hydraulic pump plus heavy system A demand **may** cause A/P A to disengage.*

◆Radio Altimeters
- Provide radio altitude to respective FCCs
- With inoperative RA, A/P will disconnect two seconds after LOC and GS capture

Autothrottles (A/T)

System Overview

Provides automatic thrust control in all regimes

FMC provides A/T system with N1 limit values

Separate servo motors for each thrust lever

Manual intervention

Thrust Levers may be manually moved any time

Unless in **THR HLD** or **ARM**, when released, A/T system will reposition thrust lever toward commanded mode setting

EEC Modes

A/Ts controlled with EECs **NORM** or **ALTN**

Both EECS recommended to be in same mode (**NORM** or **ALTN**) for minimizing thrust lever separation

Engagement and Disengagement

Engaged if armed switch in ARM and one of the following buttons is depressed or active

Thrust Lever—**TO/GA** Switch

MCP— ♦**N1** ♦**VNAV** ♦**SPEED** ♦**ALT HOLD**
♦**ALT ACQ** ♦**V/S** ♦**GS** capture ♦**LVL CHG**

Note: *Advancing thrust levers while simultaneously pressing* **TO/GA** *may lead to invalid thrust setting or* **TO/GA** *not engaging properly*

Disengaged for any of the following

♦A/T Arm Switch moved to **OFF**
♦A/T Disengage Switch (either) pressed
♦A/T system fault is detected
♦Two seconds after landing touchdown
♦Significant thrust difference along with control wheel roll input of 10° or more, and flap position up through 10
♦Autopilot / Autothrottle Indicators

Autopilot (A/P) Disengage Light—**RED**
♦Flashes & tone sounds when A/P disengages (Reset by pressing a Disengage Light or A/P Disengage Switch
♦Steady for any of following conditions:
♦♦**ALT ACQ** mode inhibited during A/P go-around if stabilizer not trimmed (single A/P operation)
♦♦Automatic ground system tests fail
♦♦Stab out of trim < 800' RA (dual channel approach)
♦ **AMBER**—Steady—Test Switch held in position 1
Flashing—A/P automatically reverts to CWS pitch or roll while in CMD. (Resets by pressing either light or selecting another mode)

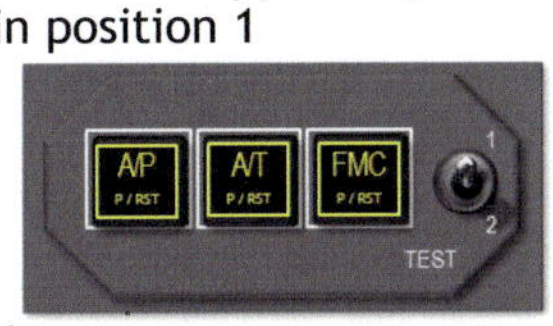

Autothrottle (A/T) Disengage Light—
RED—Flashing—A/T disengaged

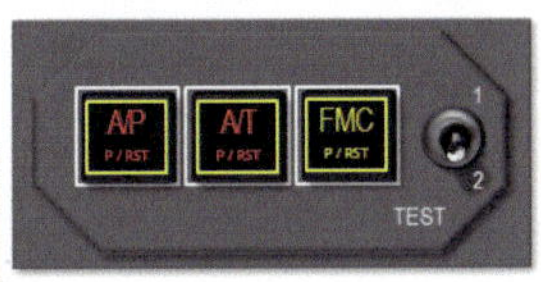

Disengage Light Test (**TEST**) Switch
TEST 1—All three disengage lights — **AMBER**
TEST 2—**A/P** & **A/T** Lights **RED**
FMC Alert Light **AMBER**

Autopilot Flight Director System (AFDS)

Flight Directors (F/Ds)

Displays

Turning on **L** or **R F/D** switch causes display of command steering and pitch bars on respective side
If no pitch or roll mode engaged, respective bar does not appear

F/D Mode and AFDS

F/Ds operate in same mode as AFDS *except:*
Takeoff mode is F/D only
Dual F/D guidance is available for single engine operations
F/D has no landing flare ability—F/D command bars disappear at 50 feet RA on ILS approach

Master FCC

Indicated by respective **MA** light
With neither A/P engaged in **CMD** mode, first F/D turned on selects master FCC
When selecting an A/P:
The first selected A/P causes the related FCC to be the master, regardless of which F/D was engaged first
Selecting the other A/P switches the FCC Master to the side of the most recently selected A/P
The Master FCC can be switched in this way without deselecting the F/Ds by selecting the A/P on the side desired for FCC master
The FCC master can also be switched by turning off the FCC that is currently the master and re-engaging it
If the other FCC was already **ON**, it becomes master FCC—It then stays master when first one is turned back on

Independent F/D Operation

Normal condition—With **both F/D switches on, both CA** and **F/O F/Ds** are driven by **master FCC**
Independent F/D Control—Each F/D is controlled directly from **respective FCC**
Conditions for independent control:
Neither A/P engaged in **CMD** mode ***and***
Both F/D switches are on ***and*** any of the following
APP Mode engaged with LOC and G/S captured
GA Mode engaged and below 400' RA
TO Mode engaged and below 400' RA
Indication—both L and R MA lights illuminated

Generator Loss

F/D will switch to electrically powered side automatically in certain conditions

Capability armed only

TO or G/A
Dual F/D APP mode below 800'

Indications

MA light illuminates on side of good generator
MA Light extinguishes on slaved side

AFDS Annunciations

Status—displayed above attitude display on outboard display unit (DU)

Possible indications:
CMD—one or both A/Ps engaged
FD—FD is on and A/P either off or in CWS
CWS-P or **CWS-R**—Pitch or Roll mode CWS engaged

AFDS Flight Mode Annunciations

Three columns, L to R ARP
Autothrottle
Roll
Pitch

Autothrottle Modes ..... See ASM, Autothrottle System 4.20.3

Pitch Mode Notes (selected highlights)

Takeoff mode sequence of commands
0-60 knots ....10° nose down
60 knots through liftoff to initial climb ....15° nose up
After "sufficient climb rate is acquired" ....Pitch to maintain MCP speed plus 20 knots
Go Around Mode
Activated when **TO/GA** button(s) pressed with the following conditions
Inflight <2000' RA ***or***
>2000' with flaps extended (out of up)
Not in takeoff mode
F/D(s) either **ON *or* OFF**
When engaged
F/Ds command roll to maintain ground track
F/Ds command pitch to 15° nose up
After reaching "programmed rate of climb"
Pitch commands target airspeed for each flap setting
Based on maximum takeoff weight

Roll Mode Notes (selected highlights)

LNAV Armed takeoff
If armed, LNAV engages at 50' RA
AFDS intercepts and tracks the active route if:
Within 3NM of the active route segment (regardless of actual heading)
Greater than 3NM from active route segment
Must be on a heading which will intercept the active route at less than 90° heading change
Must intercept the active route before the active waypoint
VOR/LOC—AFDS tracks selected VOR or localizer inbound front course bearing

Automatic Flight Operations

Takeoff and Climb

See Takeoff & Climb notes, *page 20* of this Study Guide

Flare and Landing Highlights

Flare cue in HUD—starts at ≈90' RA
- If stabilized, indicates flare commencement at ≈50'RA

A/T Reduction—begins at ≈27' RA so as to reach idle at touchdown

A/T disconnection occurs ≈2 seconds after touchdown

Single Engine F/D Go-Around

Commenced with press of either TO/GA switch
- FMA Display indications
  - ◆F/D holds current track; FMA Roll window ....Blank
  - ◆Pitch mode—TO/GA ....**TO / GA**
  - ◆Thrust—(single push) advances toward reduced N1 to produce 1,000-2,000 fpm climb **N1**
- Target speed is displayed on
  - IAS/Mach Window ***and*** PFD & HUD airspeed cursor
- Initial Pitch commanded—13° nose up
- As climb rate increases, pitch commanded maintains a target speed, displayed on airspeed cursor
- Target speed is based on engine failure conditions—If engine fails:
  - Prior to G/A engagement—target speed is the selected MCP speed
  - After G/A engagement—target speed depends on whether 10 seconds have elapsed since G/A engagement:
    - Prior to 10 seconds—MCP selected approach speed becomes target speed
    - Over 10 seconds *AND* engine failure airspeed is
      - Within 5 knots of G/A engagement speed, airspeed existing at go-around engagement becomes target speed
      - More than 5 knots above G/A engagement speed, current airspeed becomes target speed

Other Target Speed Notes
- The F/D target speed is displayed
  - On IAS / Mach display
  - On the airspeed cursor
- Target speed is never less than V2 speed based on flap position unless in windshear conditions
- F/D commanded acceleration cannot occur until higher speed is selected on the MCP IAS/Mach display

Windshear Logic

Low Altitude—windshear encountered on F/D Takeoff or G/A
- Initial command—maintain V2+20 until V/S decreases to +600 fpm
- V/S < 600 fpm—Pitch commands 15° nose up
- If airspeed continues to decrease—15° nose up until stick shaker is reached
- Intermittent stick shaker activation commanded until airspeed increases
- Above actions are reversed as appropriate and windshear area is departed

On Final Approach
- F/D continues to command pitch to maintain ILS LOC and G/S
- Pilot intervention required, or F/D may allow airspeed decrease without regard to stick shaker

Command Speed Limiting and Reversion

Independent system—not connected to stall warning / stick shaker or mach warning

Speeds protected
- VMO/MMO
- Wing Flap placards
- Landing gear placard
- Minimum speed
  - Based on AOA
  - Is ≈1.3 VS for current flap configuration

Commanded speed cannot be selected above or below (as appropriate) protected speeds

If speed is selected in exceedance of protected speed
- AFS allows acceleration /deceleration to just short of limit, then commands the limit speed
- Overspeed or underspeed symbol appears in MCP IAS/Mach display
- Pitch or thrust (as appropriate for engaged mode) attempts to hold the limit speed
- A speed 15 knots above limit must be selected to remove underspeed limiting symbol

Speed Symbols

Underspeed—flashing **"A"**

Overspeed—flashing **"8"**

Communications

Voice Recorder

Capacity
- Four independent channels
- Records flight deck audio for 30 minutes
- Recordings older than 30 minutes (120 minutes for 3DM and subsequent) automatically erased

Channel Assignments
- 1—Flight deck area microphone
- 3—Record individual comm. Box output for pilots and observer

Erasing Tape
- Airplane on the ground
- Parking Brake Set
- AC Power Connected
- Hold **ERASE** for 5 Seconds

TEST Switch
- 3AA-3DL—Monitor indicator rising to green band indicates successful test
  - One "bump" for each of the 4 channels
- 3DM & Subsequent—
  - Momentary **STATUS** light indications good test

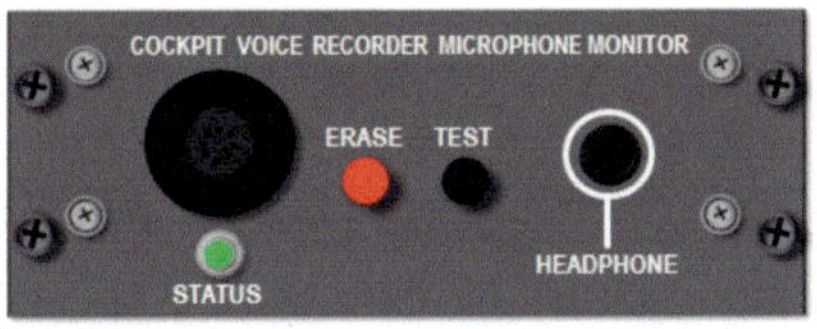

Flight Recorder
- Described in Flight Instruments chapter —*see page 53*

Radio Panel Notes

Radio Tuning Panel Off Switch
- Turns all panel functions **OFF** / **ON**

Panel Fail Modes
- **INOP**—The selected radio is not available
- **PANEL FAIL**—The radio tuning panel has failed
- Selected frequencies may be displayed in the frequency indicator when the radio is not available

Service Interphone (Overhead Panel)

Service Interphone Switch
- **OFF**—External jacks are deactivated
  - Calls between cockpit & F/As still work
  - Nose gear plug in for crew chief still functions
- **ON**—Adds external jacks to service interphone system

Control Stand Aft End Plug-In
- With microphone installed, used to communicate with flight attendant stations:
- With **SERVICE INTERPHONE** Switch **ON**, also used to communicate with any external jack location
- Bypasses audio control panels
- Allows hand mic to be used to connect to PA and Interphone

ACARS & CPDLC

Both Systems are accessed through CDUs

ACARS
- ARINC Communications Addressing and Reporting System

See ASM 5.20.14, AOM 21d.2.1

CPDLC
- Controller to Pilot Data Link Communications

See AOM 21d.3.1

Audio Control Panels (ACP)

ACPs for each pilot and observer (3)

Electronics in E/E compartment

Degraded Audio System Operation
- If ACP or E/E electronics fail, degraded mode
- **ALT / NORM** switch on ACP selects mode
- Volumes preset
- Hand mic inop

ACP Capabilities in ALT
- Captain ....T/R on VHF 1 only
- F/O ....T/R on VHF 2 only
- Observer ....T/R on VHF 1 only
- No other ACP capabilities available in ALT
  - PA, Interphone, etc
  - Aft control stand plug in ports still work

Flight Interphone System

Independent network

Allows uninterrupted communication between cockpit *and*
- ♦Flight Attendants
- ♦Ground Personnel

Separate from Service Interphone

Oxygen Mask and Communications

If left oxygen mask door is open
- Oxygen mask mic is enabled
- Boom mic is disabled

To reset for normal operations
- Left oxygen mask door closed
- **RESET/TEST** Switch pressed
- Also disables oxygen mask mic

Broadband System Power Switch

OFF—removes power from the Broadband System

ON—provides power to the Broadband System.

Electrical Systems

Overview

Three main Subsystems

AC Power System
- Primary Power System
- Power Sources
  - Two Integrated Drive Generators (IDG)
    - AC Power Output
      - Three-phase, 115 Volts
      - 400 Cycle
    - Each IDG powers its own respective AC bus
    - Each IDG can power opposite side
      - Essential & non-essential loads
  - APU generator—Can power both AC transfer busses on ground or in flight

DC Power System
- Power Sources
  - 3 Transformer-rectifier units (TR)
  - Battery
  - Battery charger
    - (incorporates another TR unit)
- Backup Power for AC & DC standby system

Standby Power System
- Ports power from one bus system to the other

Two Basic Principles of Operation
- ♦No paralleling of AC sources of power
- ♦A new source of power being connected to a transfer bus automatically disconnects an existing source

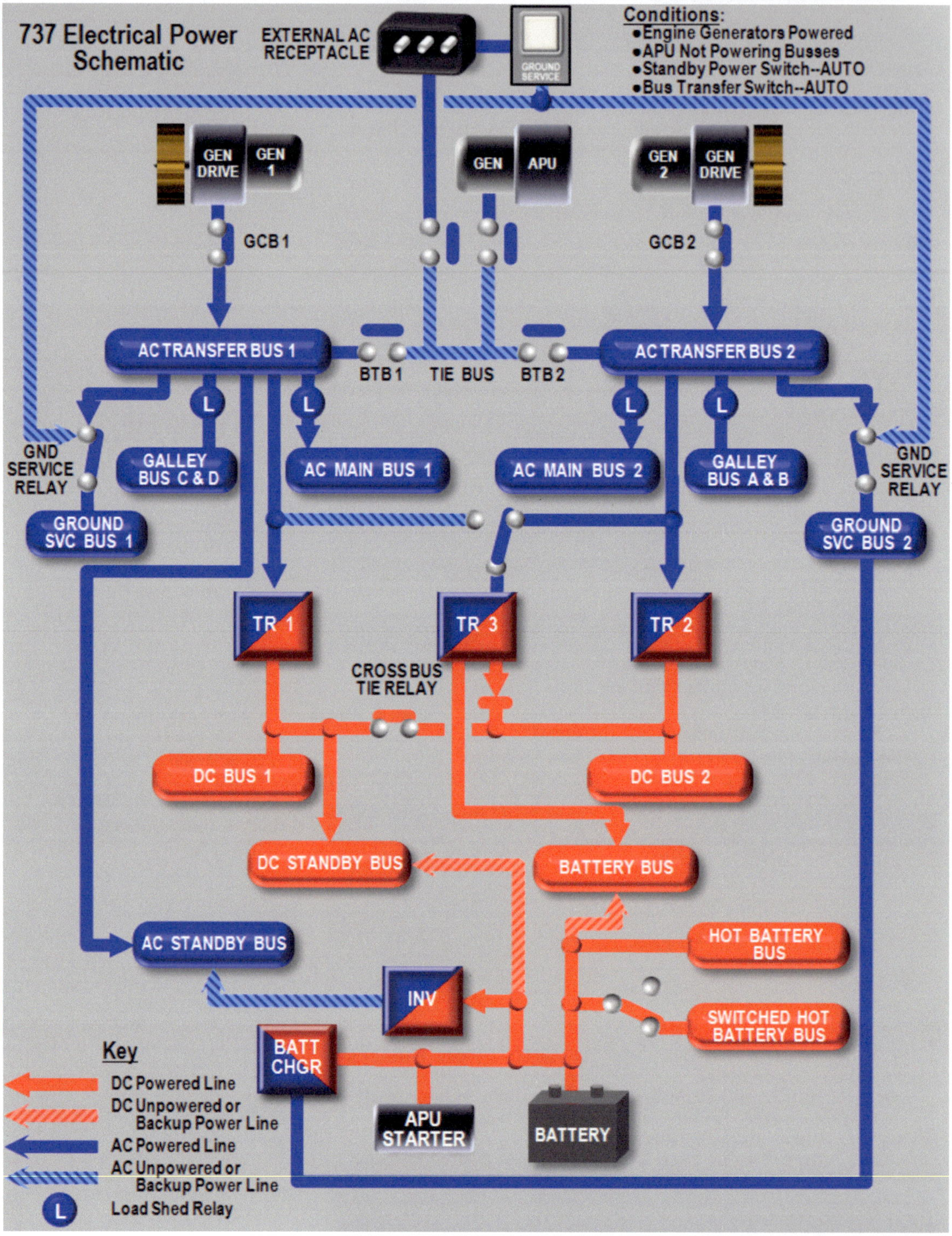

Electrical Power Generation

Engine Integrated Drive Generators (IDG)
- Maintain constant speed throughout engine operating range
- Integral electro-optical disconnect device provides IDG mechanical isolation

APU Generator
- Can supply power to both AC Transfer busses on ground or in flight
- As sole power source, can meet power requirements for
  - All ground conditions
  - Most flight conditions

External Ground Power
- Switch on overhead panel
- Provides power to all aircraft systems
- External power receptacle panel status lights allow ground crew to see if power is connected

AC Power System

AC Power System Components

Transfer Bus
- Transfer Bus 1 provides power to AC Standby Bus
- AC Source Interruption
  - Occurs when a source is disconnected or fails
  - Bus Tie Breakers used to allow power to flow through Tie Bus to unpowered AC Transfer Bus

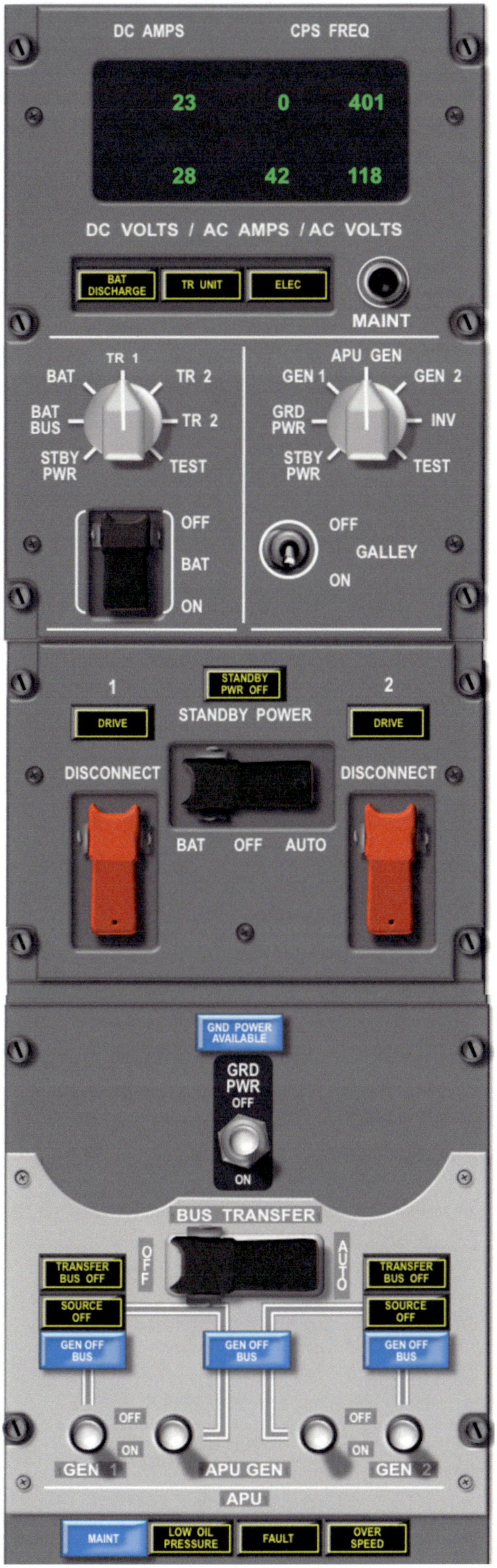

Main Busses

AC Main Bus 1 powered by AC Transfer Bus 1
AC Main Bus 2 powered by AC Transfer Bus 2
Term "Main Busses" is somewhat misleading
- On 737, Transfer Busses carry most of the critical items, not main busses
- Main busses are shed in high load conditions
- Key items powered by main busses include
  - ◆Heating elements to various probes
  - ◆Recirculation fans

Galley Busses (2)

Galley busses A & B are on AC Transfer Bus 2
Galley busses C & D are on AC Transfer Bus 1

Ground Service Bus

Switch on Forward Flight Attendant's panel
System powers AC ground service busses for
- Utility outlets
- Cabin lighting
- Battery charger

Does NOT power other busses
Switch functions are overridden when both AC transfer busses are powered
Ground Service Switch
- Located on forward F/A panel
- Momentary push to connect/ disconnect
- Illuminates (white) when power is connected

NORMAL AC Indications

Voltage .... 115 ±5 Volts
Frequency .... 400 ±10 CPS
Amps .... Vary with actual load

Powering of Transfer Busses

On Ground

Selecting external power or APU generator—both transfer busses powered
Whichever source is selected last replaces previous source
- No paralleling
- Cannot power one bus with APU and the other with external power

Powering from A/C generators—momentary push of either Generator Switch to **ON**
- Powers onside transfer bus
- Other bus remains powered by APU / external power

In Flight

Each generator powers its onside transfer bus
If one generator stops supplying power
- BTBs automatically close to allow alternate side generator to power both transfer busses

APU can power either or both transfer busses through BTBs

Automatic Generator On-Line Feature

Activated when airplane takes off with
- ◆APU powering both transfer busses ***and***
- ◆IDGs not connected to busses ***and***
- ◆APU is shut down or fails

Activation connects each generator to its related transfer bus
Can occur only once per flight, and only *in* flight

Ground Power and APU Connection to Transfer Busses

Ground Power Connection
- **GRD PWR** switch **ON**—Ground Power connects to both transfer busses
- If engine or APU generator(s) are powering busses, displaces them on all busses when **GRD PWR** switch is selected **ON** (details below)

APU Connection
- With APU operating, selecting either **APU GEN** switch **ON**...
- Connects APU power to both transfer busses

No split bus power possible (APU to one transfer bus, external power to the other)

Bus Tie System

If existing power source for one bus fails, ***and*** **BUS TRANS** Switch is in **AUTO**
- ◆Operating power source will power both busses
- ◆Works with airplane generators or APU generator (not ground power)

Remaining powered transfer bus automatically powers the unpowered bus through the BTBs

Load Shedding

Engine Generators

If one generator fails, loads shed incrementally as needed
Uses load sensing logic to automatically shed as needed
- ♦Galleys ♦Main Bus ♦IFE Bus

Restoration—Occurs automatically as load allows
Manual reset can be attempted by cycling buses with switches **OFF** & **ON**
- ♦**GALLEY** Bus ♦**CAB/UTIL** Bus

APU Generator

In flight, if APU is only electrical power source, Galley, main & IFE busses may be shed
If load still above design limits, both IFE busses are shed
Restoration as aboveAC Switches

Generator Drive DISCONNECT Switches

Disconnects IDG if
- Electrical power is available ***and***
- Engine Start Lever is in **IDLE**

IDG cannot be reconnected in the air

STANDBY POWER Switch

AUTO (Guarded)—In flight, or on the ground:
- AC Transfer Busses powered:
  - AC Standby Bus is powered by AC Transfer Bus 1
  - DC Standby Bus is powered by TR1, TR2 and TR3
- With loss of all AC power
  - AC Standby Bus is powered by battery through Static Inverter
  - DC Standby Bus is powered by battery
  - Battery Bus is powered by battery

OFF (center position) -
- **STANDBY PWR OFF** Light illuminates
  - AC Standby Bus, Static Inverter, and DC Standby Bus are not powered

BAT (unguarded position) -
- AC Standby Bus is powered by battery through Static Inverter
- DC Standby Bus and Battery Bus are powered directly by battery

Galley & Utility Bus Switches

<u>**GALLEY** Power Switch</u>—supplies power to galleys when AC transfer buses are powered

<u>**CAB/UTIL** Power Switch</u> (3DM and subsequent)
- Powers the following galley & cabin equipment
  - ♦All 115V AC galley busses
  - ♦Recirculation fans
  - ♦Lavatory water heaters
  - ♦Door area heaters
  - ♦Potable water compressor
  - ♦Drain mast heaters
  - ♦115V AC shaver outlets when installed

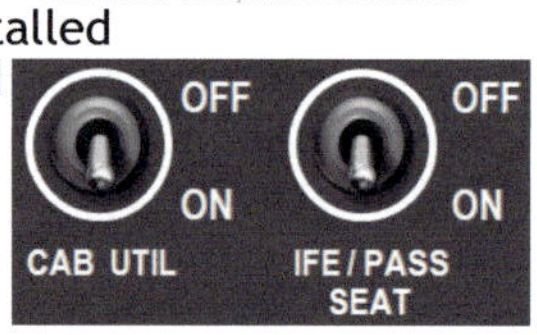

<u>**IFE/SEAT PASS** Power Switch</u> (3DM and subsequent)
- Powers the following components of passenger seats, in-flight entertainment systems, and other including:
  - ♦115V AC audio & video entertainment equipment
  - ♦Cabin telephone equipment and FAX machine
  - ♦28V DC video equipment
  - ♦Passenger seat electronic outlets

AC System Status Lights

Generator Drive (DRIVE) Lights—Low IDG oil pressure caused by one of the following:................. **MC** **ELEC**

- ♦IDG failure
- ♦Engine shutdown
- ♦IDG automatic disconnect due to high oil temperature
- ♦IDG disconnected through Generator Drive **DISCONNECT** Switch

STANDBY PWR OFF Light—One or more of the following are unpowered: .... **MC** **ELEC**

- ♦AC Standby Bus
- ♦DC Standby Bus
- ♦Battery Bus

Ground Power & Bus Switching Panel

Generator Bus Lights

GRD POWER AVAILABLE—Power Source available, meets airplane power quality standards

APU GEN OFF BUS—APU Running, not powering a bus

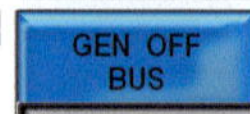

GEN OFF BUS—IDG not powering related transfer bus

Status Lights

TRANSFER BUS OFF—Related transfer bus unpowered........................ **MC** **ELEC**

SOURCE OFF Lights................. **MC** **ELEC**
- No source has been manually selected to power the related transfer bus, ***or***
- The manually selected source has been disconnected
- If a source was previously selected to power the opposite transfer bus, both transfer busses are powered

Bus Switches

Ground Power Switch—3 position switch, spring-loaded to neutral
- **OFF**—Disconnects ground power from AC Transfer Busses
- **ON**—If momentarily moved to **ON** position and ground power is available:
  - Removes previously connected power from AC Transfer Busses
  - Connects ground power to AC Transfer Busses if power quality is correct

GEN (Generator) Switches

Three position switch, spring-loaded neutral
- **OFF**—Disconnects IDG from related AC Transfer Bus by opening generator breaker
- **ON**—Connects IDG to related AC Transfer Bus by disconnecting previous power source and closing generator circuit breaker

BUS TRANSFER Switch

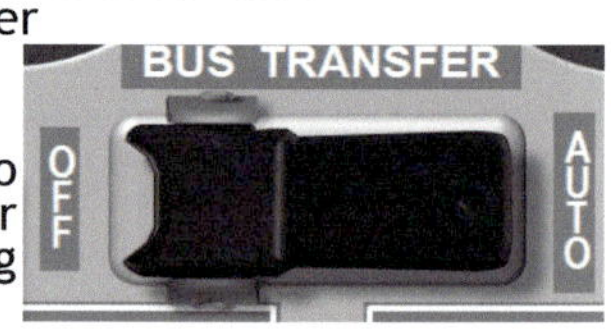

AUTO (Guarded)
- BTBs operate automatically to maintain power to AC Transfer Busses from any operating generator or external power
- DC Cross Tie Relay
  - automatically provides normal or isolated operation as required

OFF—Isolates AC Transfer Bus 1 from AC Transfer Bus 2 if one IDG is supplying power to both AC Transfer Busses
- DC Cross Tie Relay opens to isolate DC Bus 1 from DC Bus 2

APU GEN—Three position switches

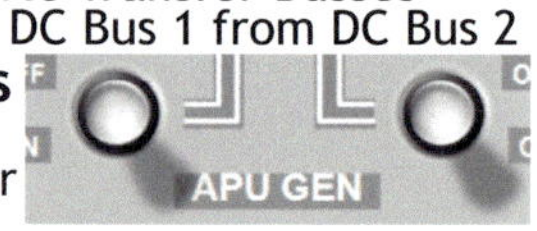

Spring-loaded to neutral

OFF—APU powering both AC Transfer Busses
- Moving a <u>single</u> **APU GEN** Switch to **OFF** illuminates related **SOURCE OFF** Light
  - APU continues to power AC Transfer Busses
- Moving other **APU GEN** Switch to **OFF**
  - ♦Disconnects APU generator from Tie Bus
  - ♦Removes APU power from AC Transfer Busses
- APU generator powering one AC Transfer Bus & IDG powering one AC Transfer Bus,
- Moving related **APU GEN** Switch to **OFF**
  - ♦Disconnects APU generator from Tie Bus and AC Transfer Bus
  - ♦IDG powers AC Transfer Busses

ON—<u>Neither AC Transfer Bus powered</u> by IDG
- Moving a single **APU GEN** Switch to **ON**:
  - ♦Connects both AC Transfer Busses to the APU generator
  - ♦Disconnects external power, if connected
  - ♦Opposite **SOURCE OFF** Light illuminates until the other **APU GEN** Switch is moved to **ON**

ON—<u>Both AC Transfer Busses powered</u> by IDGs
- Moving a single **APU GEN** Switch **ON**:
  - ♦Powers the related AC Transfer Bus from the APU generator
  - ♦Other AC Transfer Bus continues to receive power from the IDG.

GROUND SERVICE Switch

Momentary push-button switch
Provides manual control of Ground Service Busses
Enables servicing airplane using external power without activating AC Transfer Busses
ON—Connects external power to Ground Service Busses
OFF—Disconnects external power from Ground Service Busses

DC Power System

DC Power Sources

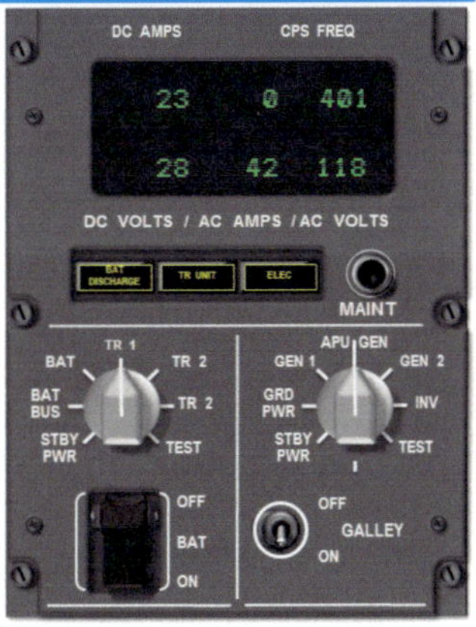

Three Transformer-Rectifiers (TR)

Convert 115 V AC to 28 V DC
TR1 powered by AC Transfer Bus 1
TR2 Powered by AC Transfer Bus 2
TR3 powered normally by AC transfer bus 2, alternately can be powered by AC Transfer bus 1

Battery

Always supplies power to Hot Battery Bus
When selected on, supplies power to
- Battery Bus
- DC Standby Bus
- Switched Hot Battery Bus
- If AC Standby bus is not powered by AC Transfer Bus 1
  - Standby Inverter is powered
  - AC Standby Bus is powered through Standby Inverter

With no other power source available, can supply power for a minimum of 30 minutes

Battery Charger Transformer Rectifier

Powered through Ground Service Bus 2
When battery is fully charged, reverts to constant voltage power generation mode for
- Hot Battery Bus
- Switched Hot Battery Bus

If Tr3 fails, powers Battery Bus
With loss of AC Transfer Bus 1 *or* the DC Bus 1, powers AC and DC Standby Busses

Battery Warning System

Alerts personnel outside A/C that aircraft is on battery power
Alert criteria
- ♦Aircraft on ground
- ♦Battery **ON** with AC power removed from buses
- ♦2 Minutes have elapsed with no change in above criteria

Inhibited when
- ♦AC Power is applied ***or***
- ♦One or both start levers out of **CUTOFF**

Installed on **3NX – 3VM**
To be retrofitted on **3DM – 3NW**

Cross Bus Tie Relay

Normally closed, connecting DC Bus 1 & 2
Automatically opens to isolate DC Bus 1 from DC Bus 2 when:
- ♦Bus Transfer Switch positioned to **OFF** ***or***
- ♦At glide slope capture during a flight director or autopilot ILS approach
  - Isolates the DC Busses during approach
  - Prevents a single failure from affecting both navigation receivers and flight control computers

Normal Operations

TR1 & TR2 power output is used to share the load of:
- DC Bus 1
- DC Bus 2
- DC Standby Bus

TR3 Powers Battery Bus
- Backup source for TR1 & TR2 busses

DC System Status Lights

BAT DISCHARGE—With **BAT** switch **ON**, excessive battery discharge detected.......... **MC ELEC**

TR UNIT.. **MC ELEC**
<u>Ground</u>—Any TR has failed
<u>In flight</u>—TR1 failed, *OR* TR2 and TR3 failed

ELEC—A fault exists in DC power system or standby power system .... **MC ELEC**
Operates only with airplane on ground

DC Switches

BATTERY Switch

OFF—Normal power available
- Removes power from
  - ♦Battery Bus ♦Switched Hot Battery Bus
- Battery only source of power—Removes power from:
  - ♦Battery Bus
  - ♦Switched Hot Battery Bus
  - ♦DC Standby Bus
  - ♦Static Inverter
  - ♦AC Standby Bus

ON (Guarded Position)
- ♦Provides power to Switched Hot Battery Bus
- ♦Energizes relays to provide automatic switching of standby electrical system to battery power with loss of normal power

Standby System

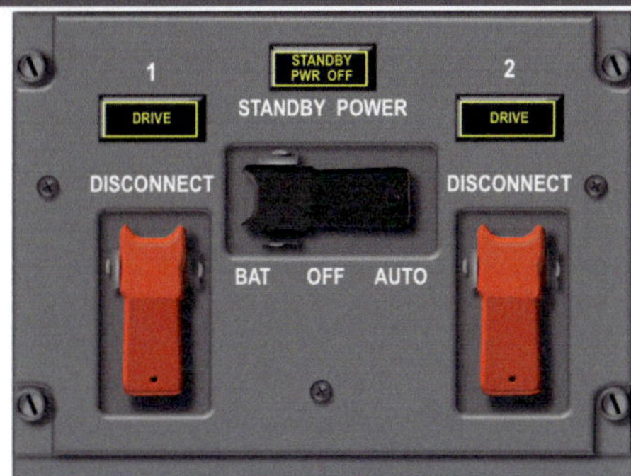

Standby Power System Includes:

- ♦Main Battery
- ♦Static Inverter
- ♦AC Standby Bus
- ♦DC Standby Bus
- ♦Battery Bus
- ♦Hot Battery Bus
- ♦Switched Hot Battery Bus

Standby Power Control

STANDBY POWER Switch

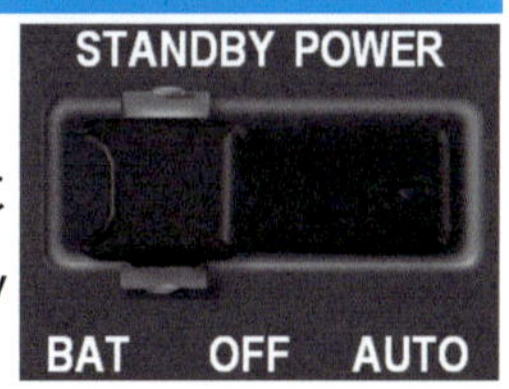

AUTO (guarded) GRound and In flight
- With AC Transfer Busses powered:
  - ♦AC Standby Bus is powered by AC Transfer Bus 1
  - ♦DC Standby Bus is powered by TR1, TR2 and TR3
- With loss of all AC power
  - ♦AC Standby Bus is powered by battery through Static Inverter
  - ♦DC Standby Bus is powered by battery
  - ♦Battery Bus is powered by battery

OFF (center unguarded)
- ♦**STANDBY PWR OFF** Light illuminates
- ♦AC Standby Bus, Static Inverter, and DC Standby Bus are not powered

BAT (unguarded)
- ♦AC Standby Bus powered by battery through Static Inverter
- ♦DC Standby Bus and Battery Bus powered directly by battery

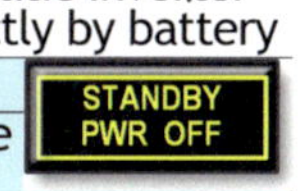

STANDBY PWR OFF Light................ **MC ELEC**
Power is <u>not</u> available to one or more of the following:
- ♦AC Standby Bus (Inverter not powered)
- ♦DC Standby Bus
- ♦Battery Bus

Engines

General

Engine Characteristics

	737-800 3AA-3PX	737-MAX 3RA-3VM
Designation	CFM56-7	LEAP 1B
Thrust	27K	28K

Type ....Dual rotor, axial flow turbofan

Compressor....................................Dual stage, N1 / N2

N1 Rotor—Low pressure compressor

N2 Rotor—High pressure compressor

Drives accessory gear box

Bleed-air-powered starter connects to N2

Engine Electronic Controls (EEC)

Dual channel

EECs regulate engine performance

Monitor thrust lever inputs and autopilot and set thrust

Alternate mode lacks some safeguards, so advancing thrust levers full forward can result in overboost

Reverse Thrust—variable idle to maximum reverse thrust

Engine Indications

Primary Engine Indications

N1 & EGT considered primary

Displayed normally on Center Upper DU

If this DU fails, automatically displayed on center lower DU if display select panel switches are in **NORMAL**

EGT displays start limit line or continuous EGT limit amber and red line as determined by engine speed

Fuel Flow

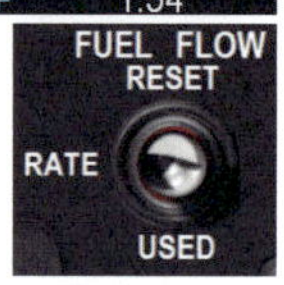

Displayed on upper DU below EGT

Fuel used displays alternately if selected

RATE—Displays engine fuel flow

USED—displays fuel used since last reset

After 10 seconds, reverts to fuel flow

RESET—Resets fuel used to zero

Displays fuel used for 1 second, then zero, then fuel flow

Engine Status Indications on N1

Thrust Reverse

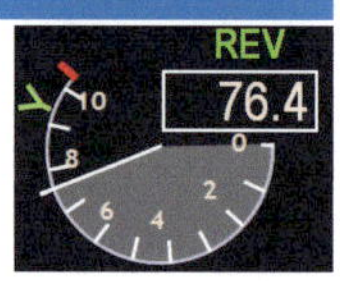

REV—Reverser moved from stowed position

REV—Reverser is deployed

Thermal Anti-Ice **(See above)**

TAI—Cowl anti-ice disagreement with switch

TAI—Cowl anti-ice valves open

Secondary Engine Indications

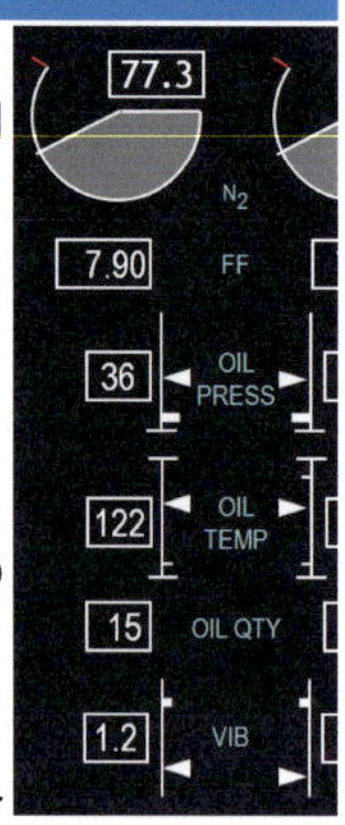

Displayed when selected on lower center DU

May be selected to Captain's or F/O's Inboard DU position

Secondary Display Indications

♦N2	♦Fuel Flow
♦Oil Pressure	♦Oil Temperature
♦Oil Quantity	♦Vibration

Automatically Displayed When:

Engine display Initial electrical power-up

A secondary engine parameter is exceeded

<u>In flight</u>, engine start lever is moved to **CUTOFF** ***or***

An engine N2 RPM is below idle

MAX Secondary display can be toggled on

No compact secondary display

N2 **MOTORING** advisory—Bowed Rotor Motoring (BRM) logic active during start

Display Colors

As conditions change, color reflects status

Normal operating ranges.............................White

Caution range.............................Amber

Exceedance range.............................Red

Abnormal Engine Indications

Oil Pressure

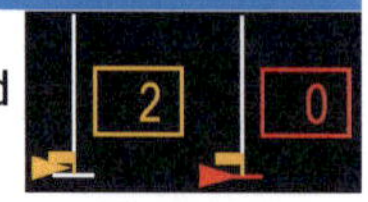

Pointer & range amber—caution range reached

Pointer & redline red at/above limit value

Actual limit values vary with engine speed

Amber not displayed below 65% N2

Oil Temperature

High oil temperature line red at operating limit

Readout changes from white to amber to red

Band amber—temperature in caution range

Pointer white, amber or red as appropriate

Oil Quantity

Displays reverse video and "LO" when quantity is low (triggering level not specified)

Zero quantity is normal for RPM below approx. 8% N2

EGT—**Displays** AMBER, RED **& ENG FAIL**

ENGINE FAIL indication on when

< Sustainable idle (50% N2)

Engine start lever in **IDLE**

Alert disappears when

Engine recovers ***or***

Start lever moved to **CUTOFF** ***or***

Fire switch pulled

Vibration Indicator

When vibration is "high," readout is in reverse video

Crossbleed Start Indication

With low RPM, X-BLD displays above N2 when airspeed too low for windmill start

With compact display, XB is shown

Empty Box

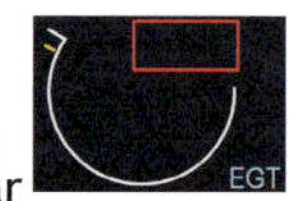

Indicates exceedance condition occurred in flight

Box will clear if engine is started, and reappear after shutdown until maintenance clears data

Crew Alerts

Light Indications—upper engine display

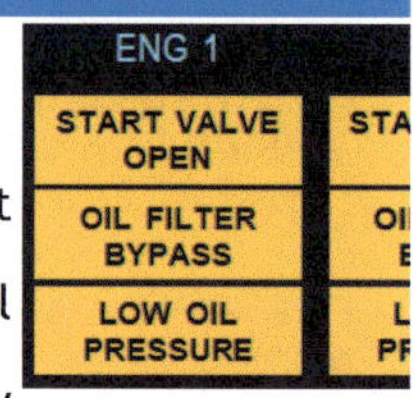

START VALVE OPEN—Respective valve open

Steady—Normal indication during start

Blinking—uncommanded opening of start valve

OIL FILTER BYPASS—Impending bypass of oil supply filter

LOW OIL PRESSURE—Oil pressure at or below red line

MAX Additional alerts

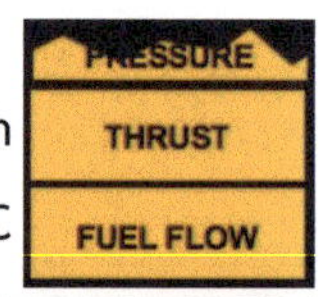

THRUST—Compares commanded thrust with actual

FUEL FLOW—Disagreement between FMC predicted and EEC sensed flow rates

For all three alerts

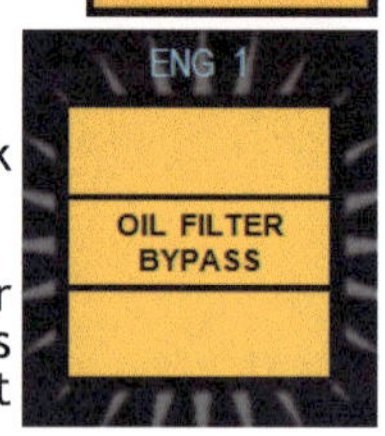

♦Annunciated Light illuminates steady

♦Boxes around all three lights and blank boxes flash for 10 seconds, then

♦Only annunciated light remains on steady

<u>During inhibited parameters</u> (below), box for only the triggered indication illuminates steady rather than flashing (others not illuminated) <u>when</u>:

<u>Takeoff</u>—Inhibit <u>starts</u> at 80 knots and <u>continues until</u>

400' RA ***or***

30 seconds after reaching 80 knots

<u>Landing</u>—Inhibit <u>starts</u> as A/C descends through 200' RA and <u>continues until</u>

30 seconds after touchdown

N1 Set Function

Allows manual control of N1 bug
Normal control is through N1 FMS function
- Outer knob selects function
- Inner knob sets N1

Compact Display

Combines primary and secondary instruments into one display
- All indications are digital other than N1 and EGT

Colors change to amber or red as above if parameters exceeded

Normally—not used
- If lower secondary instrument display functional, blanking is preferred
- Allows screen to pop up automatically if exceedance occurs rather than having to crosscheck the display
- N2 display shows XB during crossbleed starts

Engine Sub-Systems

Engine Panel

Reverser lights—On when (ATIS): MC ENG

- ♦**A**-*Auto restow circuit* has been activated
- ♦**T**-*Thrust Reverser*: One or more thrust reverser sleeves not in commanded state
- ♦**I**-*Isolation Valve OR* Thrust Reverser Control Valve not in commanded position
- ♦**S**-*Sync*: Failure in synchronization shaft lock circuitry

MAX Additional Reverser Lights

- **REVERSER COMMAND**—Reverser commanded to deploy in flight
- **REVERSER AIR/GND**—Reverser Air/Ground logic protection is lost
- **REVERSER LIMITED**—Reverse thrust for respective engine unavailable
  - Replaces -800 **REVERSER** lights on panel above

Engine control lights—On when: MC ENG

- Engine control system NOT dispatchable due to system fault(s)
- Light operates ONLY on ground with engine operating and:
  - ♦A/S below 80 knots prior to takeoff ***or***
  - ♦Approx. 30 seconds after touchdown

Electronic Engine Control (EECs)

General Function

- EECs provide data for some indications after they are powered
- Battery Start with no AC available
  - EECs powered after N2 reaches 15%
  - Before engine start switch placed in **GRD**, raw data shown
  - Instruments affected
    - ♦ N1
    - ♦ N2
    - ♦ Oil Quantity
    - ♦ Engine Vibration
  - Prior to Battery **ON**
    - 3 Instruments—self-generating display capability
      - ♦ N1
      - ♦ N2
      - ♦ Oil Quantity
- Normal Start with AC
  - EEC power up begins during engine start when engine start switch is placed in **GRD**
  - After start switch in **GRD** with AC power, EECs provide data for instruments above

EEC Modes

<u>Normal Mode</u>
- **ON** in view, **ALTN** blank
- Calculates N1 thrust settings
- Derated takeoff uses less than full thrust lever
- EEC uses actual flight conditions & bleed air demand
- EEC drives power to achieve thrust rating selected
  - -800 ....27, 26, 24 or 22K pounds per engine
  - MAX ....TO (28K), TO1 (-10%), TO2, (-20%)

<u>Takeoff Bump</u> Thrust
- Authorized only at SNA
- Applies only to takeoff thrust, not climb or other settings
- "Provides thrust above normal maximum takeoff thrust"

<u>Alternate EEC Mode</u> .... MC ENG

- Soft Alternate Mode
  - If required signals not available, EEC automatically switches to soft alternate
  - Indications—**ALTN** and **ON** both displayed
  - Uses last sensed flight conditions
  - As conditions change, exceedances are possible
    - Climb or descent
    - Temperature, etc.

- Hard Alternate is selected instead
  - Indicated by **ALTN** with **ON** out of view
  - Assures thrust at least equal to what Normal Mode would provide

Structural Limit Protection
- N1 and N2 overspeed—EEC provides protection, in both Normal and Alternate modes
- EGT Limit—<u>**no**</u> EEC protection in any mode
  - Pilots must monitor
  - Post flight-exceedance displays regardless of EEC mode

Idle Power—Three modes, set by EEC

<u>Ground</u> Idle—lowest thrust idle setting
<u>Flight Minimum</u> Idle—selected for most phases of flight
<u>Approach</u> Idle
- Higher thrust N1 & N2 than flight minimum idle at same altitude and temperature
- Improves <u>engine acceleration</u> for go-arounds *and/or*
  - Provides <u>additional bleed air</u> production (anti ice on)
- Selected by EEC in flight if
  - Flaps are in landing configuration ***or***
  - Engine anti-ice is **ON** for either engine
- Maintained until after touchdown

Signal Loss
- If EEC stops receiving flap or anti-ice signals due to a fault, approach idle is activated at 15,000 feet MSL

MAX Thrust Control Malfunction Accommodation (TCMA)

Protects against unacceptable high asymmetric thrust on the ground, including an RTO or landing
<u>If detected</u>, EEC shuts off fuel to affected engine
ENG FAIL Alert illuminates when N2 falls below 50%

MAX Electronic Overspeed System (EOS)

Protects against engine structural design limit exceedance
<u>If detected</u>, EEC shuts off fuel to affected engine.
ENG FAIL Alert illuminates when N2 falls below 50%

Engine Fuel System

Fuel Path from Tanks to Engine

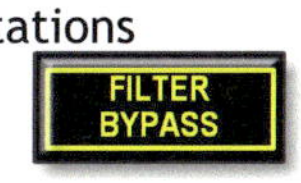

- Fuel Boost Pumps
- <u>Fuel Spar Shutoff Valve</u>—engine mounting wing stations
- Fuel-oil heat exchangers
- Fuel filter with bypass
  - If filter saturates, bypass occurs
  - **FILTER BYPASS** light alerts pilots (Fuel Panel, Overhead console)............................. MC FUEL
- Second stage engine fuel pump
- Hydro Mechanical Unit (HMU)—(EEC meters fuel through HMU)
- <u>Engine Fuel Shutoff Valve</u> is between HMU and engine

Two Fuel valves (above) start and end the path

- Both open when Start Lever is in **IDLE** ***<u>and</u>***
  - Fire Switch is down
- Fuel flow measured downstream of engine fuel shutoff valve
- Status lights on Fuel panel signal when valves are open by going OUT (dark)
  - **ENG VALVE CLOSED**
  - **SPAR VALVE CLOSED**

Engine Oil System

Oil circulates under pressure to lubricate
- Engine bearings
- Accessory gearbox

Oil Indications on secondary engine display unit
- ♦ Oil Quantity Indicator
- ♦ Oil Temperature Indicator
- ♦ Oil Pressure Indicator
- ♦ **LOW OIL PRESSURE** Alert

Oil is pressurized by engine driven oil pump

Sensors for indicators and warning light located downstream of the oil pump, prior to engine lubrication

Engine Driven Scavenge Pumps return oil to the oil tank

Oil Supply Filter is downstream of scavenge pumps
- If filter is saturated, oil bypasses the filter
- **OIL FILTER BYPASS** alerts if this occurs

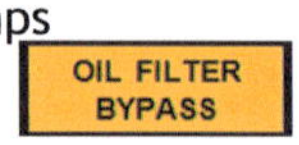

Main Engine Oil Cooler uses fuel to control cool and control oil temperature

Engine Start System

Power Requirements

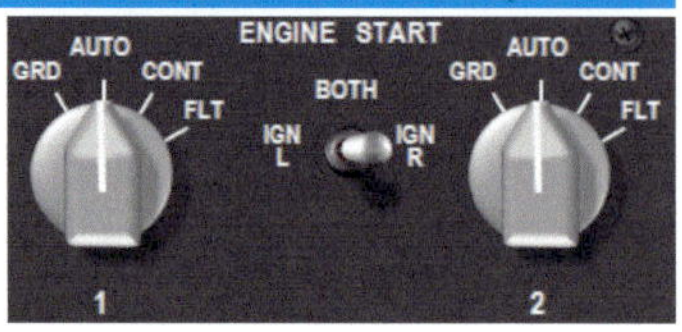

Bleed air system pressurized air powers the starter motor
- Possible Sources
  - ♦APU
  - ♦External air
  - ♦The other operating engine

Electrical power
- Possible sources
  - ♦Battery
  - ♦External Power
  - ♦APU
  - ♦The other operating engine

ENGINE START Switches

GRD—Opens Start Valve ***and***
- ♦Closes Engine Bleed Valve
- ♦when Engine Start Lever is moved to **IDLE**
  - Ground starts—arms selected igniter(s)
  - Inflight starts—arms both igniters
- ♦Releases to **AUTO** at start valve cutout

AUTO—Ignition normally off, , for flameout protection:
- Both igniters activate when Engine Start Lever in **IDLE** ***and***
  - ♦Uncommanded rapid decrease in N2 occurs, ***or***
  - ♦N2 is between 65% and 50%, ***or***
  - ♦In flight—N2 is between idle and 5%
- Selected igniters activate when Engine is running, ***and***
  - ♦Flaps are not up below 18000 feet altitude, ***or***
  - ♦Engine anti-ice is selected to **ON**

CONT—Selected igniters activated when
- ♦Engine is operating ***and***
- ♦Engine Start Lever is in **IDLE**
- Both igniters activated inflight when
  - N2 is below idle ***and***
  - Engine Start Lever is in **IDLE**

FLT—Both igniters activated when Engine Start Lever in **IDLE**

ENGINE START Levers

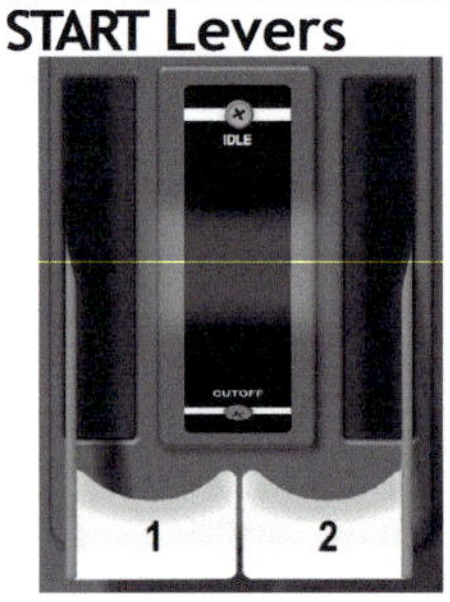

MAX **CAUTION** Do not apply rotational force when moving the engine start levers

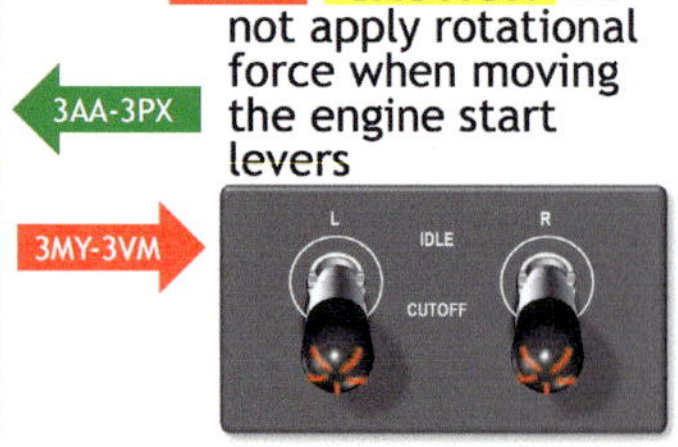

IDLE -
- ♦Energizes ignition system through EEC
- ♦Electrically opens spar fuel shutoff valve in the wing leading edge outboard of the pylon
- ♦Electrically opens engine-mounted fuel shutoff valve via the EEC

CUTOFF -
- ♦Closes both spar and engine fuel shutoff valves
- ♦De-energizes ignition system

Ignition Select Switch
- **IGN L**—Selects the left igniter for use on both engines
- **BOTH**—Selects both igniters for use on both engines
- **IGN R**—Selects the right igniter for use on both engines

Start Sequence
Engine Start Switch—GRD position
- Uses battery power to...
  - Close the Engine Bleed Air Valve ***and***
  - Open the start valve to
  - Allow pressure to rotate the starter

When the start valve opens, START VALVE OPEN alert displays on upper display unit

Starter rotates N2 compressor through the accessory drive gear system

Engine accelerates to the recommended speed—25% N2 ***or*** Max motoring

Engine Start Lever is moved to the **IDLE** position
- Opens the fuel valves on the wing spar and engine
- EEC then supplies fuel and ignition to the combustor
- Fuel then ignites

Initial fuel flow indications lag actual fuel flow by approximately two seconds—May cause EGT rise before fuel flow indication

At starter cutout speed (approximately 56% N2), power is removed from the Start Switch holding solenoid

The Engine Start Switch returns to **AUTO**

Engine Bleed Air Valve returns to the selected position, Start valve closes

Auto-Relight
Causes both igniters to be activate

Detected when uncommanded rapid decrease in N2 sensed, OR when N2 is below idle RPM

MAX Bowed Roter Motoring (BRM)

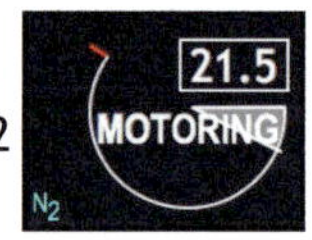

EEC Function to correct for thermal deformation
- Heat following shutdown can deform N1 & N2 rotor shafts

Active for 12-60 seconds on the ground only:
- After Engine Start Switch moved to the GND ***and***
- Before Engine Start Lever moved to IDLE

During BRM, **MOTORING** displays on N2 gauge between 18% & 24% N2

MAX Additional Tests
TCMA and EOS functions

When N2 accelerates to recommended value (25% N2 or max motoring), engine start lever is moved to IDLE

Indications:
- Fuel flow ....0 (Zero)
- Engine fuel shutoff valve ....Opens and closes repeatedly
- **ENG VALVE CLOSED** light ......ON, bright blue during logic test
- ....Extinguishes when the test complete

After test completes, start sequence continues
- Fuel valves on the wing spar and engine open
- EEC to supply fuel and ignition to the combustor

Inflight Starting

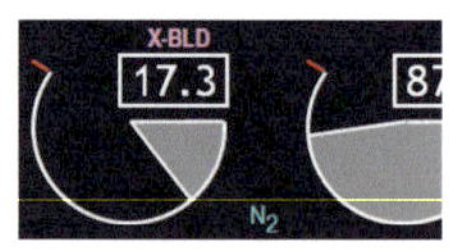

Two sources for engine rotation—windmill or crossbleed

If A/S too low for windmilling *and* N2<50%
- X-BLD indication on N2
- XB in compressed mode

Oil Quantity may read low
- Due to low rotation speed of oil scavenge pump
- Should be accurate after restart

Abnormal Start Protection—Ground Starts Only
EEC monitors engine parameters to detect 3 conditions
- ♦Impending hot starts
  - Detected by a rapid rise in EGT *OR*
    - EGT approaching the start limit,
  - Resulting warning: white box surrounding the EGT digital readout flashes
    - Box resets when Start Lever is moved to **CUTOFF** ***or*** Engine reaches idle N2

◆EGT Start Limit exceedences
If EGT exceeds the starting limit:
EGT box and dial display turn red
EEC automatically turns off the ignition *AND*
Shuts off fuel to the engine
Alert terminates and display returns to white when EGT drops below the start limit
Following shutdown of both engines
EGT box turns red
Reminds crew of the exceedance

◆Wet Starts
Occurs if the EGT does not rise after the Start Lever is moved to **IDLE**
If detected, the EEC turns off the ignition ***and***
Shuts off fuel to the engine 15 seconds after the Start Lever is moved to **IDLE**

Protection features do *NOT* function during inflight starts

Reasons to Abort Start Attempt

Start lever **CUTOFF**—No N1 Rotation
Response—Start Switch to **AUTO**

Start Lever in **IDLE**
◆No EGT Increase in 10 Seconds
◆Little or no increase in N1 or N2
◆EGT Quickly approaches or goes above start limit
Response—
Start lever to **CUTOFF**
Motor engine 60 seconds
Start Switch to **AUTO**

Stabilized in **IDLE**—No oil pressure
Response—Start Lever to **CUTOFF**

For any of these, the immediate action applies:
Engine Start Lever (affected engine)..................**CUTOFF**

Engine Warm up

Thermal stabilization period of three minutes required:
After start, before takeoff
After termination of reverse thrust before shutdown

Oil temperature requirements

Must be verified ≥ 31°C before setting takeoff thrust
Running engines 3 minutes with OAT ≥ 0°C allows required temperature stabilization

Starter Limits

Duty Cycle Limits AOM 18.10.3	737-800 3AA-3PX	737-MAX 3RA-3VM
Multiple consecutive start attempts permitted	Applies to both	
Minimum 10 seconds between start attempts		
Max allowable starter engaged time per attempt	2 Minutes	3 Minutes

Extended Engine Motoring AOM 18.10.3	737-800 3AA-3PX	737-MAX 3RA-3VM
Maximum time—1st & 2nd extended start attempts	15 Minutes	5 Minutes
Minimum time between first two attempts	2 Minutes	5 Minutes
Maximum time for 3rd and subsequent attempts	5 Minutes	
Minimum time between attempts, after second	10 Minutes	
Motoring Procedure: see AOM 21f.1.8, **"On-Gate Engine Motoring Procedure"**	X	

Engine Thermal Stabilization

Time recommended for warmup

From stable idle to application of T/O thrust..........3 Minutes
3RA-3VM Engine oil temperature must be ..at or above 31°C ....AOM 5.2.4

Time recommended for cool down

No longer required..AOM 1.2.4

Thrust Reversers

Hydraulically actuated

Engine 1—System A Hydraulic system
Engine 2—System B hydraulic system
Alternate source—Standby Hydraulic System
Deploys & Retracts slower if this source is called on
Some thrust asymmetry can be expected

Deployment possible

Either radio altimeter senses less than 10' AGL ***and***
Thrust Levers in Idle position (mechanical restriction)

Indications

—Reverser moved from stowed position
—Reverser is deployed
Amber **REV** indication on upper DU—On when either reverser sleeve moves from stowed position
Green **REV** indication when reverser reaches deployed condition
Reverse lever can then be moved to detent 2
"When necessary," after detent 2 reached, reverse lever can be pulled to maximum reverse thrust
Amber **REVERSER** light on Aft Overhead Panel

On when thrust reverser commanded to stow
Extinguishes when isolation valve closes (Approx. 10 seconds later)

Abnormal Deployment/Stowing

If **REVERSER** light on for over 12 seconds, indication of malfunction .... MC ENG
Pause in movement of reverse thrust levers past detent 1 toward stow position
May cause same indications
800 A pause of approximately **16** seconds engages an electro-mechanical lock
MAX A pause of approximately **8** seconds engages an electro-mechanical lock
May prevent reverser sleeve movement
Cycling thrust levers may clear fault and restore normal operation

Note: *Additional detail on reverser lights in* *Engine Panel description**, this section*

Auxiliary Power Unit

APU Operation

Functions
- Provides electrical power for engine generator backup
- Provides air for starting and air conditioning & pressurization

Envelope
- Start & operation ....Up to A/C maximum certified altitude
- Both AC Transfer Busses and bleed air ....≤10,000' MSL
- Bleed air alone...≤17,000' MSL
- Electrical power alone ....≤41,000' MSL

Fuel Supply
- From left side fuel manifold when AC pumps operating
- Suction fed from No. 1 tank when no pumps operating
- Heated when APU running to prevent fuel icing

Engine & Cooling Air
- Air intake through inlet on AC right aft fuselage
- Exhaust through muffler out aft end of fuselage
- APU cooling air
  - Enters through inlet above exhaust outlet
  - Cools oil through oil cooler
  - Vents through APU exhaust

Electrical Requirements
- In order to operate:
  - ◆APU Fire Switch on Overheat/Fire Panel must be IN
  - ◆APU Fire Control Switch on APU Ground Control Panel must be IN
  - ◆Battery Switch must be **ON**
- Starting Power, in order of preference if available
  - ◆No. 1 Transfer Bus
  - ◆Aircraft Battery
- APU Electronic Control Unit
  - Requires battery **ON** to power
  - If battery is turned off, no power to control unit, and APU shuts down automatically

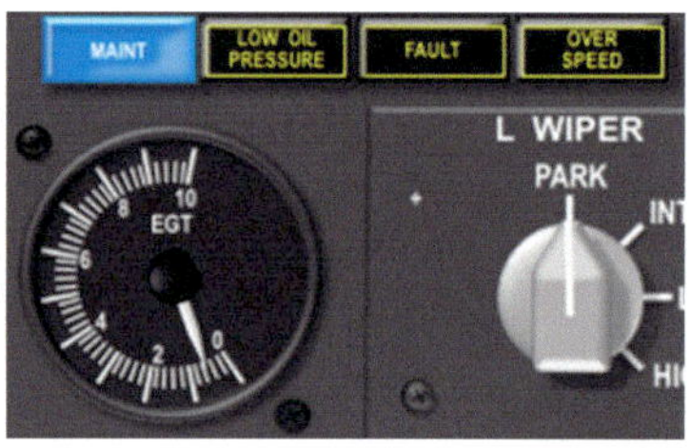

APU Starting
- Sequence is automatic
  - EGT may cycle from 0-1100°C momentarily
  - **LOW OIL PRESSURE** light may cycle several times
  - **APU GEN OFF BUS** should come on by end of start cycle
- Cycle Duration
  - May last up to 120 seconds
  - Automatic shutdown for EGT exceedance
- Bleed Air Use—APU must be run 2 full minutes before opening APU bleed air valve

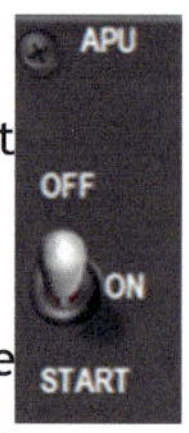

Note: 3RA- 3VM If external power is not used for APU start:
- Wait at least 30 seconds between closing the BATTERY switch guard and starting the APU
- Allows for completion of DPC self-test. ....AOM 3.2.1

MAX —**EGT Gage** is removed from APU Panel

APU Warning Lights

MAINT Light—APU maintenance problem exists
- APU may be operated
- Light is disarmed when APU Switch **OFF**
- Most common cause is (reportedly) low oil quantity
- MAX MAINT light removed

MAX **DOOR Light**—Door not in commanded position
- Three APU door positions—
  - ◆ Fully open
  - ◆ Closed
  - ◆ Partially open
- Light on if:
  - APU is shut down and APU door is open (flight or ground)
  - In flight - APU on and door is not in flight (partial) position
  - On ground - APU on and door is not in ground position.

APU LOW OIL PRESSURE Light .... MC APU
- On during start until APU oil pressure is normal
- After start—low oil pressure resulting in automatic shutdown
- If light **ON** when APU Switch is placed to **OFF**, extinguishes after 5 minutes
- Light disarms when APU Switch is **OFF**

APU FAULT Light—indicates:............ MC APU
- A malfunction caused the start to fail and initiate an automatic shutdown
- If light is **ON** when APU Switch is placed to **OFF**, extinguishes after 5 minutes
- Disarmed when APU Switch is **OFF**

APU OVER SPEED Light .... MC APU
- Indicates either:
  - ◆APU RPM limit was exceeded; automatic shutdown *or*
  - ◆Overspeed shutdown protection feature failed self-test during normal APU shutdown
- If light illuminates when APU Switch is placed to **OFF**, extinguishes after 5 minutes
- Light disarms when APU Switch is **OFF**

APU Shutdown

Cooldown 1 minute after deselecting bleed air

If BAT switch is ON, If APU was operating
- wait approximately 5 minutes after the **APU GEN OFF BUS** Light extinguishes before placing the Battery Switch **OFF**. ....AOM
- Operate the APU for one full minute with no bleed air load prior to shutdown.
  - Recommended to extend the service life of the APU
  - When APU Switch is moved to OFF, *time delay is met automatically*

Note: Moving the Battery (BAT) Switch to OFF prior to the completion of the three minute cool down / shutdown cycle will:
- ◆ Automatically shut down the APU
- ◆ Premature shutdown event can harm the APU
- ◆ Event is recorded in a permanent data chip
- ◆ Chip is sent to the manufacturer during overhaul and results in a charge to American Airlines.

Turning battery OFF during shutdown / cooldown cycle
- Terminates the timer ***and***
- Sets premature shutdown event in a permanent data chip
- Procedure calls for
  - If APU is operating, delay battery shut off until approximately .... 3AA-3PX 2 minutes
  - .... 3RA- 3VM 5 minutes
  - after the APU GEN OFF BUS light extinguishes
  - ....APU Shutdown – AOM 21c.1.2
  - ....Secure Checklist – AOM 14f.4.1

APU switch to OFF
- Closes APU bleed air valve
- Extinguishes APU **GEN OFF BUS** light
- Shutdown occurs automatically after 60 seconds
- When speed decreases, fuel valve and inlet door close
  - If fuel valve does not close, **FAULT** light on after ≈30 seconds

Electronic Control Unit

Monitors and controls the APU

Manages automatic shutdowns described above

Electrical load monitoring and shedding
- Some loads shed if speed or EGT exceeds acceptable levels
- During engine start, electrical loads shed before air
- When running and acceptable levels exceeded, inlet guide vanes decrease bleed air load and protect electrical load if possible

APU Automatic load shedding

Inflight, with APU only source of electrical power, loads are shed in the following order:
- ALL galley busses and main busses MAY shed
- Both IFE busses

On ground
- Galley and main busses shed to maintain load within limits
- Manual restoration may be attempted

Fire Protection

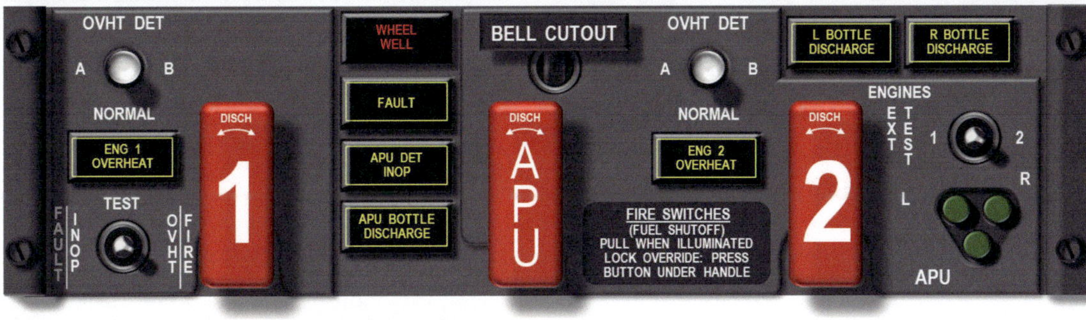

General

Fire Protection & Overheat Detection

Component	Fire Detection	Extinguishing System	Overheat
Both Engines	X	X	X
APU	X	X	
Cargo Compartments	X	X	
Lavatories	X	X	
Main Wheel Wells	X		

Power Sources

Power Sources	Fire Detection	Fire Protection
Engines, APU	Battery Bus	Hot Battery Bus
Cargo Compartments	DC Bus 1 and DC Bus 2	
Lavatories	DC Bus 1	
Trash Bin Ext. System	Self-contained battery operated	
Main Wheel Wells	No. 2 AC Transfer Bus	

Warning Lights

Master Fire Warning (FIRE WARN) Lights

Illuminated—fire warning (or system test) in
- ♦Engine ♦Main Gear Wheel Well
- ♦APU ♦Cargo compartment

Engine & APU Detection and Protection

Engine Fire & Overheat Detection

Fire & Overheat Loops

Two overheat/fire detection loops per engine
Each loop senses temperatures and activates a warning at predetermined temperature for overheat, fire
Normal ops—both loops must detect to activate warnings
Activation stops when temperatures lower to normal range
Single loop (A or B) can be selected with OVHT DET switch

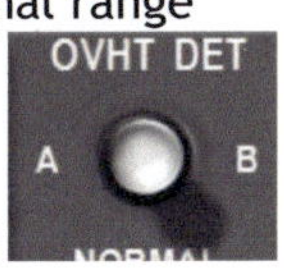

Fault Monitoring

Fault monitoring circuit monitors loops
If one loop detected bad, it is deselected automatically
Remaining single loop functions to triggers warnings
No flight deck indication occurs
If both loops fail on an engine, FAULT light results

Engine Overheat Indications—4 Lights

ENG 1 (or 2) OVERHEAT (1)..... MC (2) OVHT/DET (1)

Engine Fire Indications—7 Lights, 1 Bell

FIRE WARN (2)............ENG 1 (or 2) OVERHEAT (1)
Related Fire Switch (1)......... MC (2) OVHT/DET (1)
Fire Bell sounds 
3MY-Subsequent ....Related start lever (1) (8 lights)

Engine FAULT Lights

With Overheat Detector Switch in NORMAL
Indicates both detector loops for an engine have failed
With the overheat Detector Switch in A or B
Indicates the selected loop for an engine has failed

Note: MC & OVHT/DET *Annunciator Lights do NOT illuminate*

Engine Fire Extinguishing

Components

- ♦2 Engine fire extinguisher bottles ♦2 Fire Switches
- ♦2 BOTTLE DISCHARGED Lights ♦An EXT TEST Switch

Either or both bottles can be discharged into either engine

Engine Fire Switches

Normally locked down
Fire OR overheat warning causes unlock
Manual unlock button below each switch if needed
Pulling up causes 8 actions
Associate with light colored areas on overhead panel

Actions when Engine Fire Switch is pulled up:

- ♦**FUEL**—2.......Closes engine fuel shutoff valve and spar valve
- ♦**AIR**—1..............................Closes engine bleed air valve
  - Results: ....Lose associated wing anti ice
  - Lose associated pack air source
- ♦**ELECTRICAL**—1 ....Trips generator control relay & breaker
- ♦**HYDRAULICS**—2 ......Closes the Hydraulic Fluid Shutoff Valve
  - Engine driven Hydraulic Pump LOW PRESSURE Light is deactivated
- ♦**ENGINES**—1 ....Related engine thrust reverser deactivated
- ♦**FIRE PANEL**—2 ......Allows Fire Switch rotation for discharge
  - Arms one discharge squib on each engine fire bottle

Actions when Fire Switch is rotated:

- ♦Electrically "fires" one squib, discharging extinguishing agent into the related engine
- ♦Rotating switch other way discharges remaining bottle
- ♦L or R BOTTLE DISCHARGED Light
  - ON after a few seconds
  - Indicates the bottle has discharged

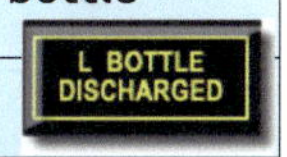

APU Fire Detection & Protection

APU Fire Detection

Single detection loop
If temperature rises to preset level—Loop signals APU fire
APU Fire switch light illuminates and stays lit until temperature below threshold

Fault Monitoring................ MC OVHT/DET

Single loop has monitoring circuit
Loop failure results in APU DET INOP light
APU Detection System is then inoperative

Fire Detection Indications

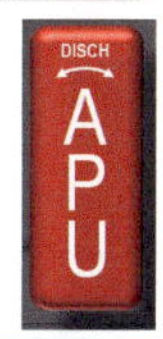

Any time:
- ♦FIRE WARN (2) ♦APU Fire Switch (1)
- ♦Fire Bell sounds ♦APU Shuts down

Ground Only—Wheel well APU Fire Warning
- ♦Horn Sounds ♦Light flashes

APU Fire Extinguishing

Components

- ♦APU Fire Switch ♦1 APU fire extinguisher bottle
- ♦EXT TEST Switch ♦APU BOTTLE DISCHARGE Light

APU Fire Switch

Normally locked down; Fire signal unlocks switch
Manual unlock button available below switch if needed

Pulling APU Fire Switch up:

- ♦Shuts down APU (Backup for automatic shutdown)
- ♦**FUEL**...Deactivates fuel solenoid; Closes APU fuel shutoff valve
- ♦**AIR**.........Closes APU bleed air valve; Closes APU air inlet door
- ♦**ELECTRICAL**......Trips APU generator control relay and breaker
- ♦**FIRE PANEL** ....Allows APU Fire Switch rotation for discharge
  - Arms APU fire extinguisher bottle squib

<u>Rotating the APU Fire Switch in either direction</u>

- ♦Electrically "fires" squib and discharges bottle
- ♦**APU BOTTLE DISCHARGE** Light **ON** after a few seconds, indicating bottle has discharged

Main Wheel Well Detection & Protection

Main Wheel Well Fire Detection

Single Fire Loop--3 Lights, Bell

Temperature threshold exceedance triggers alert

Main gear fire indications ....**WHEEL WELL** (1)

FIRE WARN (2).............Fire Bell sounds

Nose gear—no warning system

Wheel Well Fire Protection

No extinguishing system provided

QRH procedure calls for lowering gear to extinguish .QRH 8.25

Cargo Compartment Detection & Protection

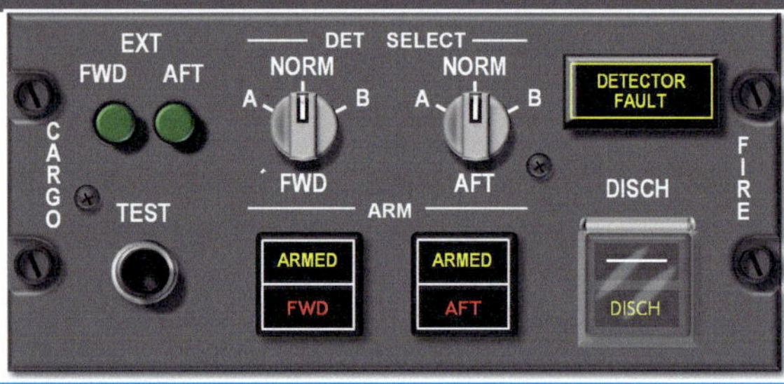

Smoke Detection

Forward and Aft Compartments

Detection is based on smoke, not temperature

Dual loop configuration

Both loops must detect smoke to trigger alert

Loops function "in the same manner as the Engine Overheat / Fire Detection Loops"

DET SEL in NORM—if a loop fails, other triggers alert

Smoke / Fire indications--3 Lights, Bell

FIRE WARN (2) ....Fire Bell sounds

FWD or AFT Cargo Fire Warning Light (1)

Cargo System DETECTOR FAULT Light

Both loops in one or both cargo compartments have failed

<u>Note</u>: MC & OVHT/DET *Annunciator Lights do NOT illuminate*

Cargo Compartment Fire Protection

Single Extinguishing Bottle

Pressing appropriate Cargo Fire Warning Light ARMs bottle for discharge to that compartment

Discharge by pressing covered **DISCH** switch

Total discharge of bottle occurs

Indications of Discharge

Cargo Fire **DISCH** light **ON**

May take up to 30 seconds to indicate bottle depletion

Duration—75 minutes, including 60 minutes inflight plus 15 minutes for go-around

Lavatory Fire Detection & Protection

Detection

All Systems Smoke Activated

<u>3AA-3DL Indications</u>

- ♦Aural Warning (in Lavatory)
- ♦Red Alarm Indicator light on lavatory smoke detector panel
- ♦Appropriate amber lavatory call light flashes
- ♦Amber lavatory **SMOKE** light on overhead panel

System resets when heat threshold no longer exceeded

Protection

Fire Extinguisher System—Heat activated (not smoke)

Beneath sink in EACH lavatory

Operates automatically

No flight deck indication of discharge

Duration of Protection— 75 minutes, to allow for

60 minutes to land

15 for missed approach, go-around landing and PAX unloading

800 75 minutes continuously metered fire suppression

MAX 75 minutes total including:

15 minutes of high rate discharge (HRD)

60 minutes of low rate discharge (LRD)

Fire System Tests

FAULT / INOP Test Detection

Tests fault detection circuits for engines and APU

Accomplished by pressing the appropriate **TEST** Switch

Extinguisher continuity tested by pressing and holding the **EXT TEST** Switch

All test indications clear when switches are released

FAULT/INOP & OVHT/FIRE Light TEST Switch

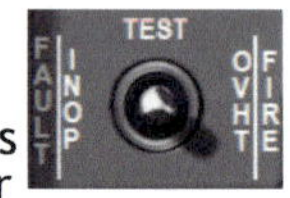

Spring-loaded to center

OVHT/FIRE—Tests overheat & fire detection loops on both engines, APU, and wheel well fire detector

<u>Indications</u>:

♦The Fire Warning Bell sounds	
♦Both Master **FIRE WARN** Lights FIRE WARN	2 Red
♦Both **MASTER CAUTION** Lights .... MC	2 Amber
♦The OVHT/DET System Light ..... OVHT/DET	1 Amber
♦Both Fire Switches illuminate	2 Red
♦The APU Fire Switch illuminates	1 Red
♦Both ENG OVERHEAT Lights illuminate	2 Amber
♦The WHEEL WELL Fire Warning Light illuminates if AC power is available	1 Red

Cockpit Total: 6 Red, 5 Amber

<u>On the ground</u>

- ♦Wheel well APU Fire Warning Horn sounds................
- ♦Wheel well APU Fire Switch flashes................ 2 Red

FAULT/INOP—Tests fault detection circuits for both engines and APU

♦Both **MASTER CAUTION** Lights .... MC	2 Amber
♦**FAULT** Annunciator Lights	1 Amber
♦**OVHT / DET** System Light .... OVHT/DET	1 Amber
♦**APU DET INOP** Light	1 Amber

Cockpit Total: 5 Amber

Extinguisher Test

Activation

EXT TEST Switch positioned to **1** or **2**

Indications

Green **EXT TEST** Lights illuminate .... 3 Green

Checks circuit continuity from squib to Fire Switch

Cargo Fire System Tests

Activation

Press and hold the Cargo Fire **TEST** Switch

Verifies extinguisher bottle squib circuit continuity

Indications

♦The Fire Warning Bell sounds..........................	
♦Both Master **FIRE WARN** Lights FIRE WARN	2 Red
♦The Extinguisher **TEST** Lights illuminate	2 Green
♦The FWD and AFT Cargo Fire Warning Lights ..... (On when all detectors in selected loop(s) respond to the fire test	2 Red
♦The Cargo Fire Bottle **DISCH** Light illuminates	1 Amber

<u>Notes</u>: *The Fire Warning BELL CUTOUT Switch can silence the fire warning bell and extinguish the Master **FIRE WARN** Lights*

- ♦ *During the Cargo Fire Test, **DETECTOR FAULT** Light illuminates if one or more detectors in the loop(s) have failed*
- ♦ *Individual detector faults only detectable by manually initiated test. **MASTER CAUTION** Light does not illuminate.*
- ♦ *After cargo fire test, up to a 4 second delay may occur to allow all applicable indications to extinguish at the same time.*

Cargo Fire Extinguisher Test

Green **EXT** Lights illuminate............... 2 Green

Verifies fire bottle discharge squib circuit continuity normal

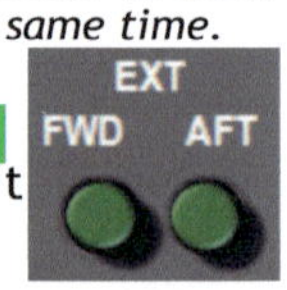

Flight Controls

Overview

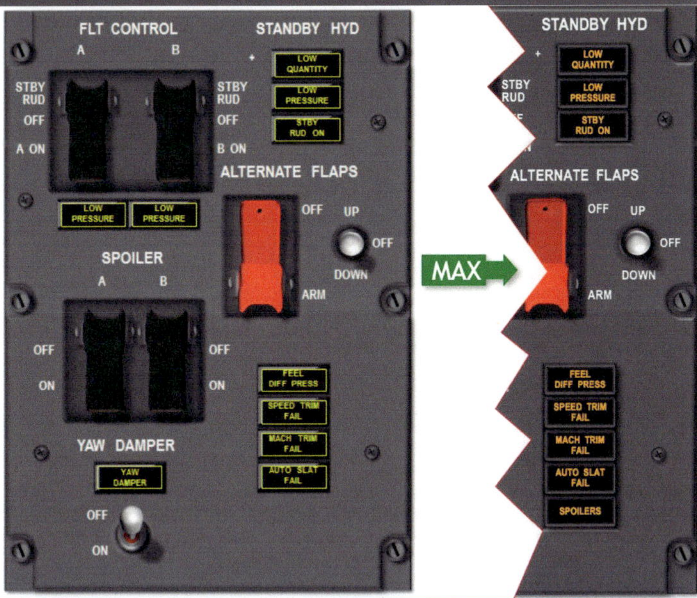

Primary Flight Controls

Consisting of:

♦Ailerons & Spoilers ♦Rudder ♦Elevators

Powered By

Dual powered actuators

Either system A or B hydraulic systems capable of powering all flight controls normally

Secondary Flight Controls

Consisting of

High lift devices

♦Leading and trailing edge flaps ♦Leading edge slats

Spoilers (flight and ground) Yaw Damper system

Powered by

High Lift devices	Hyd Sys	Backup/Function
Leading Edge Flaps & Slats	B	PTU—Extend & Retract Standby Hyd—Extend only
Trailing Edge Flaps	B	Electrical—Ext/Ret
Spoilers	A / B	Half on each System
Yaw Damper	B	Standby System

Primary Flight Controls

Roll—Ailerons & Spoilers

Control Power

Ailerons—Hydraulic systems A and B

Flight Spoilers—matched pairs on A and B systems

Positioned by yokes

Yokes are mechanically connected

Captain's yoke connected by cables to aileron actuators

F/O's yoke connected by cables to spoilers

In case of hydraulic failure

- Mechanical positioning possible
- Stiff forces due to air loads and friction

Aileron Transfer Mechanism

If one yoke jams, opposite can still control roll

Alternate yoke is then inoperative

Aileron Trim

Electrically repositions center point of yoke

If trim applied with Autopilot engaged

- Autopilot overpowers trim
- Abrupt control input on A/P disengagement
- Therefore—prohibited per LIMITATIONS

Flight Spoilers

Four on each wing (diagram, next page)

Supplement ailerons for roll responsiveness

Matched pairs on A and B system

- A or B hydraulic failure results in symmetric spoiler loss
- Remaining system powers matched powered spoiler sets

Spoilers also used as speed brakes

- Spoiler mixer controls deployment
- Allows roll combined with drag when required

Spoilers faired on with aileron down and vice versa

- Spoiler deployment starts at ≈1.7 units of yoke displacement

MAX **Spoilers**—Fly-by-wire, incorporating five features:

Maneuvering Load Alleviation (MLA)

- HIgh gross weight spoiler profile modifications
- Decreases structural loads
- Returns to commanded position when conditions allow

Landing Attitude Modifiers (LAM)

- 30-40° Flaps—Symmetric spoiler adjustment increases pitch
- 15-30° Flaps with near idle thrust—adds drag to assist in glide path acquisition

Emergency Descent Spoilers (EDS)

- Increases spoiler deployment to higher than normal position
- Armed when above FL 300 ***and*** **CABIN ALTITUDE** alert activates
- No crew action required

Elevator Jam Landing Assist System

- Improves controllability control columns become jammed
- Uses flight spoilers to make small flight path changes
- Activation: **ELEVATOR JAM LANDING ASSIST** switch manually selected ON, provided
  - ♦Actual flap position is 1° or greater ***and***
  - ♦Autopilot is disengaged.
- Results: Flight spoilers deploy to a preset position
  - ♦Control wheel steering force sensors detect forces applied to control column
  - ♦Push forward on control column causes the spoilers to extend farther, increasing the descent rate
  - ♦Pulling aft on control column causes the spoilers to retract, decreasing the descent rate
  - ♦**ASSIST ON** Light illuminates

Various logic changes to speedbrakes

MAX **SPOILERS Light**—One or more spoiler pair is inoperative

Pitch—Elevators and Stabilizer

Control Power

Elevators—hydraulic systems A and B

Stabilizer—electric motor or manual trim wheel

Flight Control A and Flight Control B shutoff switches isolate hydraulics from elevator actuators

Torque tube connects L and R control columns

With loss of hydraulics, mechanical displacement possible but difficult due to friction and air loads

Elevator Control Column Override Mechanism

With one column jammed, other can be broken free

Whichever column breaks free can provide adequate control

Other remains jammed

Flare is possible but more difficult than manual reversion

Elevator Feel System

Feel and centering unit provides simulated feel

Pressure from System A or B, whichever is higher

Airspeed inputs from elevator pitot system

If either hydraulic or pitot system fails

- Differential pressure sensed
- **FEEL DIFF PRESS** light results .... MC **FLT CONT**

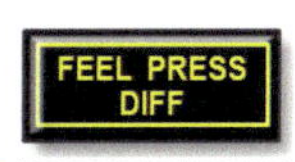

Control Yoke

- 3AA-3PX Pushing / Pulling
  - Elevators move in the appropriate direction
  - Movement opposing stabilizer trim stops electric trimming.
- 3RA-3VM
  - Pushing forward
    - Elevators move in the nose down direction
    - Interrupts trim commands from the following sources:
      - ♦Main electric trim
      - ♦Autopilot trim
      - ♦Nose up stab trim
      - ♦Speed Trim—System Speed Trim & MCAS functions
  - Pulling back
    - Elevators move in the nose up direction
    - Interrupts trim from the following sources:
      - ♦Main electric trim
      - ♦Autopilot and Speed Trim function nose down stabilizer trim commands
    - Does not interrupt MCAS nose down stabilizer trim commands.

Mach Trim System

- Provides speed stability at higher Mach numbers
- Occurs automatically above Mach .615
- Elevators are adjusted "with respect to the stabilizer"
- **MACH TRIM FAIL** Light .... MC FLT CONT
  - With annunciator system recall—single FCC channel failure
  - Without recall—mach trim system failure

Speed Trim System (STS)

- Automatically trims stabilizer
- Trims opposite speed change
- Conditions for operation
  - Airspeed between 100 KIAS and Mach 0.5
  - 10 seconds after takeoff
  - 5 Seconds following release of Trim Switches
  - Autopilot not engaged
  - Sensing of trim requirement
- **SPEED TRIM FAIL** Light—with MC Annunciator recall—single FCC channel failure.......... MC FLT CONT
  - Without recall—speed trim system failure

3AA-3PX Maneuver Characteristic Augmentation System (MCAS)

- Provides augmented pitch control *when autopilot disengaged*
- Operates to decrease high AOA when thresholds are met
- Inhibited when **SPEED TRIM FAIL** light is illuminated
- With flaps up, as AOA approaches stick shaker
  - Stabilizer nose down trim commands transition from Speed Trim function to MCAS function.
  - Trim schedules produce a predictable, increasing column force with increasing aft column displacement.
  - With column aft—
    - Column force increase is more pronounced *with*
    - Onset of EFS
- **Note**: *The Speed Trim System (Speed Trim & MCAS functions) inhibited with either:*
  - ***STAB TRIM PRI** cutout switch or*
  - ***STAB TRIM B/U** cutout switch ....are in **CUTOUT** position*
- See detailed information ....
  - ....ASM Flight Controls p. 9.20.10; 9-20-12;9-20-14

Stabilizer Trim

- Single electric motor drives stabilizer position
  - Dual speed—high when flaps extended
  - A/P uses Autopilot stabilizer trim circuit
  - Motor electrical disconnect possible through
    - **STAB TRIM MAIN ELECT** cutout switch
    - **STAB TRIM AUTOPILOT** cutout switch
    - Both on control stand
- Stabilizer Trim Cutout
  - Switches
  - Control column actuated
  - Control column forces
    - opposite trim direction also cut off trim motors
  - MAX—Switches renamed to **PRI**mary and **B/U** (backup)

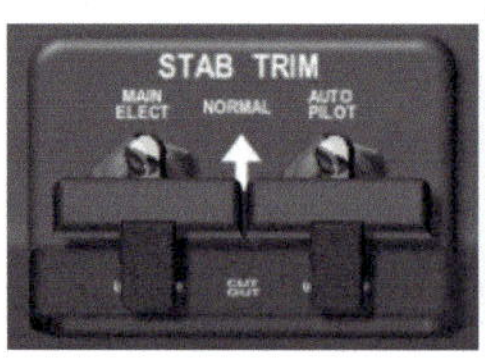

- **STAB TRIM** Override Switch
  - When positioned to OVERRIDE, electric trim can be used regardless of control column cutout switch function
- Manual Trim Wheel—can override motors or manually position trim
- Trim Limits
  - Electrical trim has limits
  - Manual trim may be required in aft CG situations
  - Electrical trim can be used to retrim toward electrical trim limit band
- Green Band—allowable takeoff range of trim

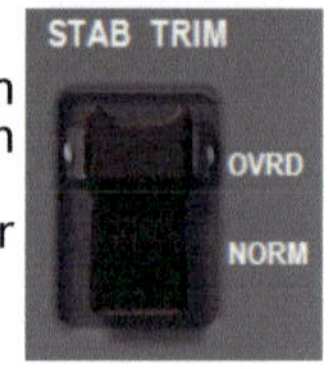

STAB OUT OF TRIM Light—See AUTOMATIC FLIGHT chapter

- Operates only with autopilot engaged (extinguished with autopilot not engaged)
- On—Autopilot not trimming stabilizer properly

Stall Identification

- Three Systems Contribute
  - Yaw Damper—During high AOA, reduces rudder inputs to decrease commanded rudder movement
  - Elevator Feel Shift (EFS) module
    - Shifts/adds additional System A pressure to push stick forward
    - Inhibited when any of the following are sensed:
      - On the ground
      - Autopilot engaged
      - RA less than 100' AGL (but if active when passing through 100' AGL, NOT inhibited)
  - Speed Trim System
    - Trims nose down approaching stall AOA
    - Allows continued trim above stick shaker AOA
- No Pilot Inputs
  - No notification that any of these have occurred
  - No pilot control of any aspect

Yaw Control—Rudder and Yaw Damper

Control Power

- Main Rudder PCU—Hydraulic systems A and B
- Standby rudder PCU—Standby Hydraulic System
- Flight Control **A** and Flight Control **B** shutoff switches isolate hydraulics from rudder actuators
- Cables mechanically connect CA and FO rudder pedals

Rudder Limiter—Rudder System Enhancement Program (RESP)

- Above approx. 135 Knots
- ♦Hydraulic system A pressure to rudder PCU limited by ≈50%
- ♦Both Hydraulic systems A and B pressures reduced within the main PCU by ≈25%.................................ASM 9.20.6
- Limits in flight authority while preserving takeoff and landing on ground authority
- Standby Hydraulic system activates on engine failure, allowing sufficient hydraulic power to operate four key components in the event of System A and/or B failure (**TRYS**)

♦**T**-*Thrust:*	Thrust Reversers
♦**R**-*Rudder:*	Rudder
♦**Y**-*Yaw Damper:*	Standby Yaw Damper
♦**S-SLATS:**	Leading Edge Flaps & Slats (extend only)

See page 61 of this Guide for related hydraulic system details

SEL 11 Rudder Differences

- Connections between CA and FO pedals are torque tubes
- Rudder limiter is 25% reduction instead of 50%
- Main Rudder PCU contains a Force Fight Monitor
  - Detects pressure difference between hyd systems A & B
  - Automatically activates the Standby Hydraulic System
  - This action allows the standby system pressure to override any rudder actuator force-fight condition between A and B hydraulics

Rudder Trim

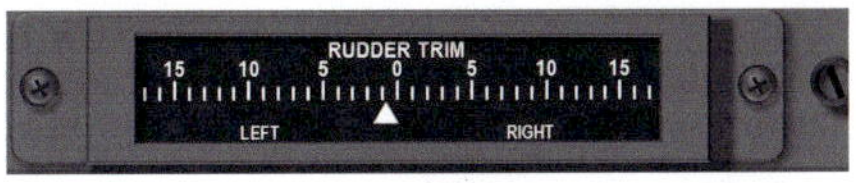

Electrically repositions the rudder feel and centering unit to adjust the neutral position

Rudder pedals are displaced proportionally

Rudder Trim Indicator displays displacement

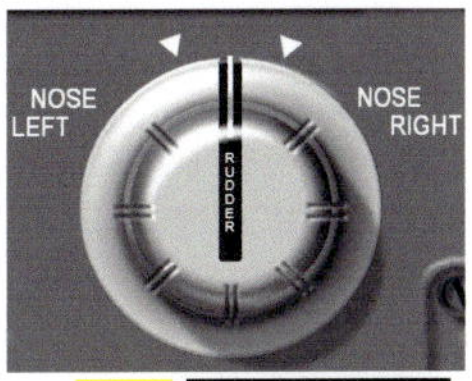

Standby Indicator Lights (All trigger MC).. MC FLT CONT

LOW QUANTITY Light

- Low quantity in standby hydraulic reservoir
- Always armed

LOW PRESSURE Light

Output pressure of standby pump is low

- Armed only when standby pump operation is selected ***or***
- Automatic standby function is activated

STBY RUD ON Light

- Standby rudder system is commanded on to pressurize the standby Rudder Power Control Unit

Yaw Damper

Consists of main and standby yaw dampers

Controlled through Stall Management / Yaw Damper computer (SMYD)

Provides damping inputs to both Main Rudder Power Control Unit (PCU) or the Standby Rudder PCU

Designed to assist in controlling
- Dutch roll
- Gust damping
- Turn coordination

Disabled automatically for any of the following:
- SMYD senses a system fault
- SMYD senses yaw damper failed to respond to a command
- **B FLT CONTROL** switch is **OFF** or **STBY RUD**

When disabled, these consequences result:
- **YAW DAMPER** switch moves to **OFF** (cannot be reset to **ON**
- **YAW DAMPER** light illuminates

Manual Reversion Impacts on Yaw Damper
- Both **FLT CONTROL** switches are placed in **STBY RUD**
- **YAW DAMPER** switch ***can*** be reset to **ON** in this condition
- Movement of control wheel signals standby rudder PCU to move rudder, assisting ailerons in turn control

YAW DAMPER Control

OFF—Disengages yaw damper

ON—Engages main yaw damper to main Rudder Power Control Unit *IF* B FLT CONTROL Switch is **ON**

Engages standby yaw damper to standby Rudder Power Control Unit *IF both* the A and B FLT CONTROL Switches are in the STBY RUD position

Light—Indicates Yaw damper not engaged .... MC FLT CONT

<u>Instructor Note</u>: *Light indicates YD computer not engaged*

Explains how light can remain off (computer still engaged) while System B loss makes YD inoperative

Secondary Flight Controls

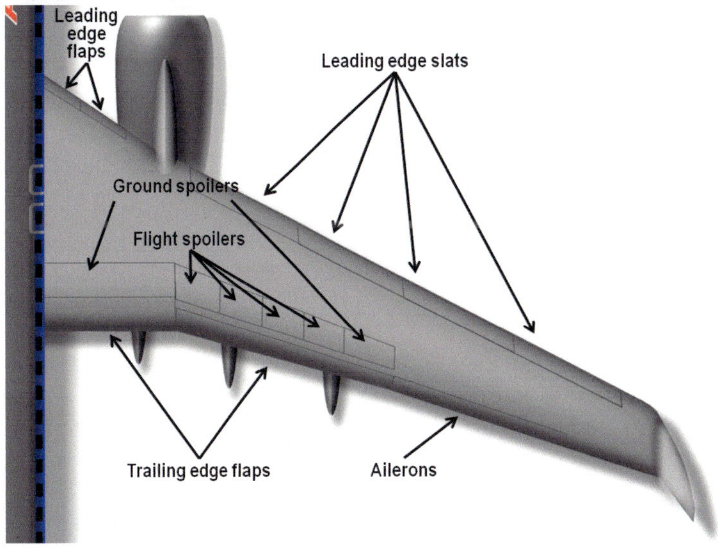

Wing Control Surfaces

Speed Brakes

Control Power

Flight spoilers also act as speed brakes—Powered in matched sets by Hyd Systems A & B

Ground Spoilers—System A Hydraulics add 4 additional wing dedicated ground spoiler panels (12 total, 6 per wing)

Activation

<u>Airborne</u>—Results in 4 panels deploying on each wing

<u>Ground</u>—Results in all 6 panels deploying on each wing

SPEED BRAKE ARMED Light

Deactivated when SPD BRK lever in DOWN position.

On—indicates valid automatic speed brake system inputs

SPEED BRAKES EXTENDED LIGHT—INDICATES FULL OR PARTIAL DEPLOYMENT ON GROUND OR IN FLIGHT

<u>IN-FLIGHT</u>—**SPEED BRAKE** Lever is beyond the **ARMED** position, ***and***
- TE flaps extended more than flaps 10 ***or***
- Radio altitude less than 800 feet

MAX <u>Additionally, for 737-MAX aircraft, in flight arming</u> occurs when—**SPEED BRAKE** Lever is beyond the **ARMED** position, ***and***
- A thrust lever is above idle for 15 seconds, ***or***
- A thrust lever is above approx. 40° for 3 seconds

<u>On the ground</u>—
- **SPEED BRAKE** Lever is in the **DOWN** detent ***and***
- Ground spoilers are not **STOWED**

<u>Note</u>—*On the ground, the* **SPEEDBRAKES EXTENDED** *light does not illuminate when Hydraulic System A pressure is less than 750 psi.*

In flight cautionary notes:

Flight spoilers increase roll rates, especially when deployed at an intermediate setting

Can be deployed past **FLIGHT DETENT** but prohibited in flight

SPEED BRAKE DO NOT ARM Light

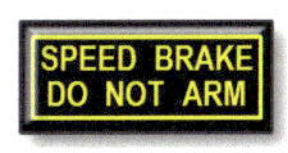

<u>On</u>—Indicates
- A/C <u>with</u> Load Alleviation System (LAS) & Flaps Up:
  - Abnormal condition of Speed Brake
- A/C <u>with</u> LAS & Flaps not UP *OR* Airplanes <u>without</u> LAS:
  - Abnormal condition ***or***
  - Test inputs to the Automatic Speed Brake System

<u>Deactivated</u> when SPEED BRAKE Lever is DOWN

MAX Relocated lights:
- **SPEED BRAKE ARMED**
- **SPEED BRAKE DO NOT ARM**
- **SPEED BRAKE EXTENDED**

Auto Speed Brake Ground Deployment

During landing, spoilers deploy automatically if all of the following occur:

- ◆**SPEED BRAKE** Lever is in the **ARMED** position
- ◆**SPEED BRAKE ARMED** Light is illuminated
- ◆Radio altitude is less than 10 feet
- ◆Landing gear strut compresses on touchdown
- ◆Compression of any landing gear strut enables the flight spoilers to deploy
- ◆Compression of the right main landing gear strut enables the ground spoilers to deploy
- ◆Both thrust levers are retarded to **IDLE**
- ◆Main landing gear wheels spin up (more than 60 kts) ***or***
- ◆If a wheel spin-up signal is *not* detected
- ◆When the Air / Ground System senses ground mode (any gear strut compresses)

Activation results in:

- ◆The **SPEED BRAKE** Lever moves to the UP position ***and***
- ◆Flight spoiler panels deploy automatically

Note: *Following an all flaps up (no flap) landing:*

- ◆*SPEED BRAKE lever will not move beyond the* **FLIGHT** *detent* **and**
- ◆*Spoilers will not fully deploy*

Speed Brake Load Alleviation System (LAS)

Installed on winglet airplanes that have unprovisioned wing (3AA - 3CL)

Designed to limit speed brake deployment airborne at high gross weights

- If Speed Brake Lever is deployed beyond 50%, it will move back to 50% position
- If speed brake lever is down, attempt to move beyond 50% is resisted by increased pressure
- Pressure can be overcome in both cases by pulling harder

Activated when all of the following occur:

- Flaps UP
- Gross weight is > 155,000 lbs
- Indicated airspeed > 320 kts

3LC & Subsequent lever stop in speedbrake mechanism

- Prevents movement beyond **FLIGHT DETENT** in flight
- Loss of electrical power eliminates protection and full movement possible

Short Field Performance

Package includes

- Sealed leading edge slats (T/O only) for improved lift
- Increased ground deployment of flight spoilers
- V Speed corrections made manually

Affects AC 3LC through 3PX

See Performance TAKEOFF 10.3, 20.4, 30.4, 40.4, 50.3

Flaps and Slats

High Lift Devices

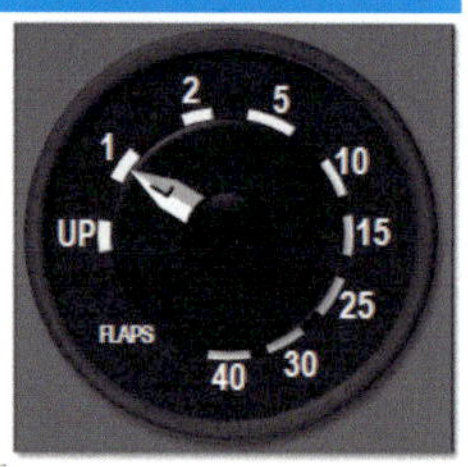

Leading Edge Devices

- 2 Inboard Flaps on each wing (4 total)
- 4 Leading edge slats on each wing (8 total)

Trailing edge flaps

- Lower settings—Increased lift ....1, 2, 5, 10 & 15
- Higher settings—Lift and drag ....15, 25, 30, & 40
- Normal landing positions............................15, 30, 40
- Authorized takeoff flap settings
  - 737-800:.....................................1, 5, 15, or 25
  - MAX:....................................1, 5, 10, 15, or 25
- Some non-normal landings and where climb performance is a factor..15
- Max altitude for deployment......................20,000'MSL
  - Excessive structural loads due to increased Mach

Control Power

Hydraulic system B is primary for both flaps & slats

Backup Sources

PTU (Power Transfer Unit—see also hydraulics, *page 61*

- Backup for the LEDs
- Activated for LEDs when System B engine driven pump pressure low *and* the flaps less than 15 but not up
- Can assist with retraction and/or autoslat operation
- Ensures normal rate movement of LEDs when operating with a number 2 engine failure

◆Standby system also provides back up for the LEDs

- Electrical Motor-driven backup for trailing edge flaps
- Purpose: Backup for LEDs following loss of system B.
- Initiated by activation of the ALTERNATE FLAPS system

Deployment Schedule—3AA - 3LB

LE Flaps are two position—retracted or fully extended

LE Slats are three position

- Retracted
- Extended (middle position, sealed)
- Full extended (slotted)

Flaps always deploy to flap actual position in degrees

Flap Position	LE Flaps	LE Slats
1, 2 or 5	Full Extend ..	Extend
>5	Full Extend ..	Full Extend

Center instrument panel lights indicate LE status

LE DEVICE annunciator panel on aft overhead panel indicates positions of individual flaps and slats

Deployment Schedule—3LC & Subsequent

Flaps always deploy to flap actual position in degrees

Flap Position	LE Flaps	LE Slats
1, 2, 5, 10, 15, 25	Full Extend ..	Extend
>25	Full Extend ..	Full Extend

LE Device Annunciator Panel

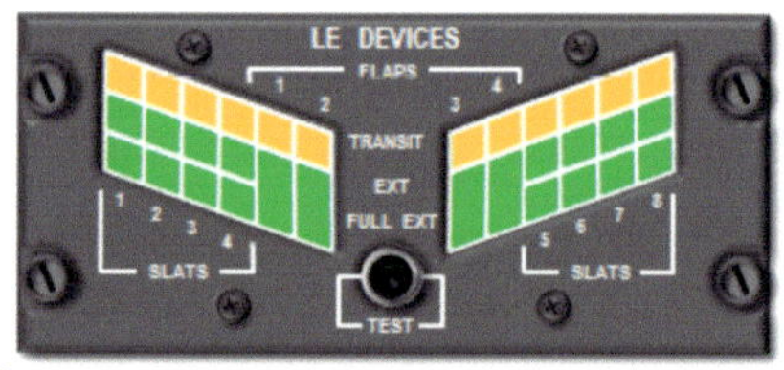

Amber—in transit

- ◆Not in programmed position for flaps
- ◆Skew exists (slats 2-7)
- ◆During alternate extension until LEs fully extended ***and*** TE flaps reach 10°

Green—

- Related LE slats Extended ***or*** Fully extended (TEs 10-40)
- Related LE flaps fully extended (TEs not UP)
- See ASM for various aircraft configuration possibilities

ASM Leading Edge Devices 9.10.8

Flap Load Relief

Flap-Slat Electronic Unit (FSEU) controls

Protects flaps from excessive forces above limit speeds

Slightly different operation for 3AA-3LB and for 3LC & subsequent

Functions only at Flaps 30 & 40 (3AA-3LB)*

Flap setting	Retract to	at Speed	and Re-Extend
40°	30°	>163 Knots	<158 Knots
30°	25°	>176 Knots	<171 Knots

Functions at flaps 10, 15, 30 and 40 (3LC & subsequent)

*Slight differences for various series versions;
See ASM Flap Load Relief 9.20.8, p. 9-20-31

Autoslats

Extends LE slats from extend (sealed) to fully extended (slotted) as AOA approaches stall

Functions at flap settings 1, 2 & 5

Activates prior to stick shaker

Return to extend position automatic as AOA decreases

Power sources

- Normal...System B
- Backup ....System A through PTU
  - (System A provides power to pressurize system B fluid)...................... *See page 60 of this Guide*

AUTO SLAT FAIL Light

- With **MC** Annunciator recall—failure of a single Stall Management Yaw Damper (SMYD) computer .... **MC** **FLT CONT**
- Without recall—auto slat system failure

Alternate Flap & Slat Extension

- Needed in the event of System B failure
- TE Flap alternate function
  - Electrical motor on AC Transfer Bus 2
  - Guarded ALTERNATE FLAPS Master Switch arms
  - **ALTERNATE FLAPS** position switch
    - Spring-loaded to neutral
    - Must be held in **DOWN** position to extend flaps
  - No asymmetry protection with electrical extension
- LE Flap & Slat alternate function
  - Same switches as above
  - Master **ALTERNATE FLAPS** switch energizes standby pump
  - Momentary **ALTERNATE FLAPS** position switch to down fully extends LE devices
  - LE devices cannot be retracted with the standby hydraulic system

Asymmetry and Skew Detection, Protection & Indication

- Asymmetry—occurs when a flap on one wing does not align with the symmetrical flap on the other wing
  - Results in small needle split corresponding to flap position
- Skew—occurs when a TE flap or LE slat panel does not operate at the same rate (L and R on the same panel) causing the panel to twist during extension or retraction
  - FSEU drives needles to obvious 15˚ split to signal skew condition
    - ....*737 Cockpit Companion*, page 186
  - No difference in procedure
  - ....See QRH FLIGHT CONTROLS 9.30 & 9.37
- FSEU monitors extension of:
  - All TE flaps
  - LE slats 2 through 7
    - 1 & 8, most outboard, not monitored
    - LE Flaps not mentioned
- TE Flap Asymmetry and Skew
  - When either is sensed, FSEU closes TE flap bypass valve, stopping extension
  - Needle split on flap position indicator shows split condition
- LE Device Improper Position or Skew
  - FSEU Monitors
  - When detected
    - **LE FLAPS TRANSIT** Light remains illuminated and *one* of the following displays on the **LE DEVICES** annunciator panel:
      - Amber **TRANSIT** Light illuminated
      - Incorrect green **EXT** or **FULL EXT** Light illuminated
      - No light illuminated
    - No skew detection of
      - Outboard slats 1 and 8,
      - LE flaps
  - During autoslat operations—Slat skew detection is inhibited

LE FLAPS TRANSIT Light

- Any LE device in transit *OR*
  - Not in programmed position with respect to TE flaps
- LE slat skew condition exists (slats 2 through 7 only)
- During alternate flap extension until LE devices are fully extended *AND* TE flaps reach flaps 10 (flaps 15, 3LC & subsequent)
- Light inhibited in flight during autoslat operation

LE FLAPS EXT Light

- All LE flaps extended ***and***
  - 3LAA - 3LB
    - TE flaps 1, 2 & 5—All LE slats intermediate position (extended)
    - TE flap positions 10 through 40—All LE devices fully extended
  - 3LC & Subsequent
    - All LE flaps extended and all LE slats in extended (intermediate) position (TE flap positions 1, 2, 5, 10, 15, and 25)
    - All LE devices fully extended (TE flap positions 30 and 40).

Uncommanded Motion Detection, Protection & Indication

- FSEU Monitors LE and TE devices
- Leading Edge Device uncommanded motion
  - No extension or retraction command present ***and***
    - 2 LE flaps move on one wing ***or***
    - 2 or more slats move on one wing
  - Results
    - FSEU shuts down the LE control
    - Amber **LE FLAPS TRANSIT** Light
  - Prevention—During cruise
    - FSEU maintains pressure on the retract lines ***and***
    - Depressurizes the extend and full extend lines
- Trailing Edge Device uncommanded motion
  - No FLAP lever or flap load relief command is present and the TE flaps:
    - Move away from the commanded position ***or***
    - Continue to move after reaching a commanded position, ***or***
    - Move in a direction opposite to that commanded
  - Results
    - FSEU shuts down TE drive unit by closing the TE Flap Bypass Valve
    - TE flap shutdown cannot be reset
    - If activated, Alternate Flap System needed to control TE flaps
  - Indication
    - Flap Position Indicator disagrees with the FLAP Lever position ***and***
    - There is no flap needle split

Flight Instruments

Note: *No attempt has been made here to completely describe the flight instrumentation displays and their connection to the FMS. These subjects are best learned using computer-aided color presentations, and are used on a day-to day basis by flight crews. Instead, key systems-related subjects have been summarized, which are easily forgotten and little used except for orals and recurrent training, and, of course, in the event of an actual malfunction. This section also serves as an overview for the transitioning crewmember.*

Primary & Standby Flight Instruments

Information Sources

Air Data Inertial Reference System (ADIRS)

- Produces flight data (position, speed, altitude & attitude)
  - Information displayed on flight display
  - Data is used by
    - Flight management computers
    - Flight controls
    - Engine controls
    - All other systems requiring inertial and air data
- Major components
  - 2 Air data inertial reference units (ADIRUs) (Described in Navigation chapter)
  - 1 Inertial system display unit (ISDU)
  - 1 Dual mode select unit (MSU)
  - 4 Air data modules (ADMs)
  - 6 Static ports
  - 3 Pitot probes
  - 2 Alpha vanes
  - 1 Total air temperature probe

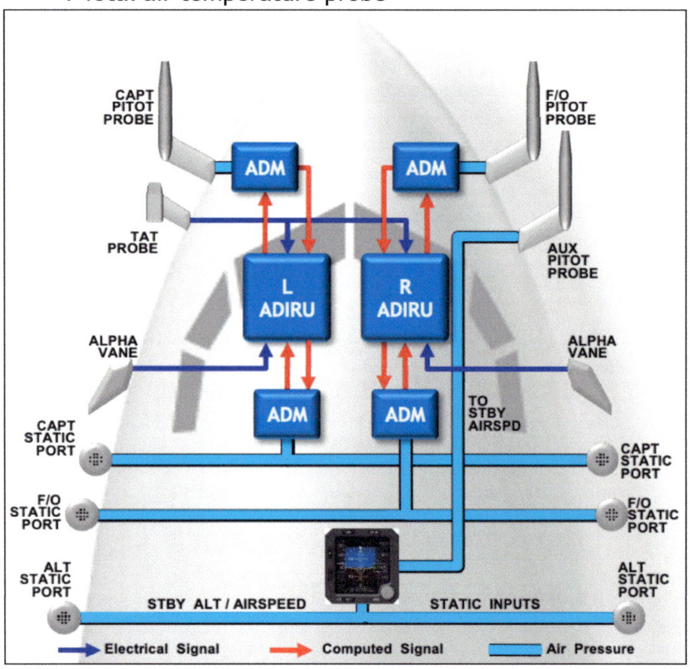

Air Data Processing

- Pitot Static System
  - 2 Pitot probes and 4 static ports feed data to ADMs
    - 2 ADMs electronically send pitot data to ADIRUs
    - 2 ADMs electronically send static data to ADIRUs
    - Processed data for use in CA and FO displays
      - L ADIRU data to CA displays
      - R ADIRU data to FO displays
  - Remaining pitot probe and 2 aux static ports
    - Feed data directly to standby altitude and A/S
    - No connection to ADIRU source pitot tubes or static ports
- Angle of Attack (AOA) System
  - 2 Alpha vanes (one on each side)—Feed data to L and R ADIRUs
  - ADIRUs feed data to PFD AOA indicators
- Total Air Temperature
  - Single probe feeds ambient temperature data to ADIRUs
  - ADIRUs compute total air temperature

Standby Flight Instruments

Standby Magnetic Compass

Standby Attitude Indicator

- Information is independent of primary attitude displays
- Power from Battery Bus—with loss of all normal AC, remains powered as long as battery power is available
- Reaches operational speed 60 seconds after power applied
- Requires 3 minutes to achieve accuracy requirements

Integrated Standby Flight Display

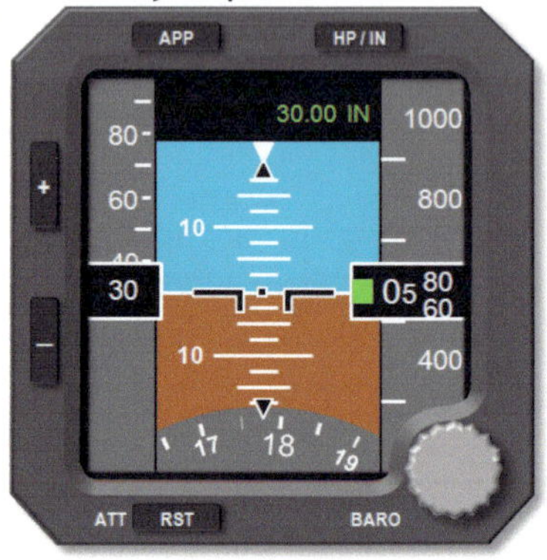

- Combines data above except magnetic compass
- Powered by Battery Bus as above
- Initialization
  - Starts 10 seconds after power applied, lasts 90 seconds
  - Interrupted if airplane movement is excessive
  - Resumes when airplane stops
- Attitude Reset Switch
  - Ground—must be activated with airplane stationary
  - In flight—must be activated in wings level, non-accelerated flight

Other Instrument Systems

Data and Voice Recording Systems

Flight Recorder

- Permanent record of selected data
- Location—Behind access door in aft cabin ceiling
- Power up and down
  - Ground—starts recording when either engine is started
  - Flight—keeps recording when either engine or APU is powering busses
- Flight Recorder Switch

  - **NORMAL**—Recorder operates as above
  - **TEST**—Powers flight recorder on the ground
  - **OFF** Light—Recorder not operating or the test invalid
    - May indicate power failure, loss of input data, or electronic malfunction

Aircraft Condition Monitoring System (ACMS)

- Digital flight data acquisition unit (DFDAU)
  - Receives signals on certain flight condition & airplane systems' operating performance
  - Converts to digital format
- Digital flight data recorder (DFDR)
  - Records data for at least the last 25 hours of operation
  - Located in aft fuselage area

Cockpit Voice Recorder

- Described in Communications chapter—*see page 36*

Head up Display (HUD) Key Parameters

Normal Takeoff

- T/O pitch limit—10° until 10' AGL to prevent tail strike
- TO/GA pitch target line displays on T/O roll > 60 knots

Low Visibility Takeoff Requirements

- Both NAV receivers tuned to same ILS frequency
- Runway length between 5500 & 13,600 feet
- Ground roll guidance engaged when heading <15° of runway heading....................... *See page 21 for requirements*

AI Approach Mode

◆HSI symbol displays until LOC course is captured ***and***
◆LOC deviation is < 0.25 degrees for 7 seconds
When criteria is met, display switches to LOC deviation scale
AI can be selected any time *except* in TO/GA or on the ground
If AI is selected prior to capturing glideslope and localizer, HUD remains in NP mode with AI mode *armed*

AII Approach Mode criteria requirements satisfied when:

◆All inputs to HUD computer are valid ***and***
◆ILS capture criteria satisfied ***and***
◆If the IRS transfer switch is in **NORMAL** or **BOTH ON L**, IRS L in **NAV** mode, ***or***
If the IRS transfer switch is in **BOTH ON R**, IRS R in **NAV** mode

AIII Approach Mode

AIII may be engaged until ≈7 seconds before 500' AGL
AIII Mode Criteria requirements satisfied when:
◆All inputs to HUD computer are valid
◆ILS capture criteria satisfied
◆IRS transfer switch in **NORMAL** with both IRSs in **NAV**
◆Both VHF NAV receivers tuned to the same ILS frequency
◆NAV switch in **NORMAL**
◆Reference glideslope angle between -2.5° & -3.0°
◆Display Source switch in **AUTO**
◆Displays Control Panel switch in **NORMAL**
◆Radio altitude > 500'
Notes: *Deviation outside these criteria prior to AIII mode selection causes reversion to NP mode*
First two items same as AII, remainder are more restrictive

Primary Flight Display (PFD) Notes

Angle of Attack Indicator

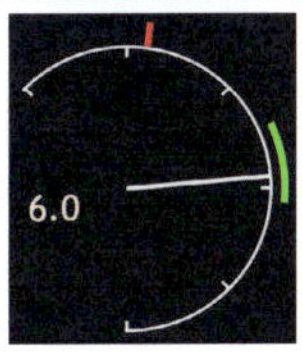

Digital AOA Readout
Indicates digital AOA value to nearest 0.2°
On ground, G/S < 80 knots—readout fixed at 0.0°
Stick Shaker Indicator (Red)
Indicates stick shaker AOA for existing flight conditions
Blank if AOA signal invalid
Analog Needle (white)—analog AOA value
Needle travel limited to -6° – +21°
Ground, G/S < 80 knots—readout fixed at 0.0° (4:30 mark)
Small differences acceptable between HUD and PFD
Approach Reference Band (green)
Shows appropriate approach AOA range for V_{REF} +5
Displayed when flaps in landing positions (15, 30, 40)
Moves with flap position
Inhibited on T/O and initial climb
Zero Degree Reference Line (white)
Indicates zero degrees angle of attack
Reference lines are displayed every 5 degrees from -5 (six o'clock position) to +20 degrees (ten-thirty position)

Instrument Landing System

Localizer and glide slope displays appear when valid signal is received

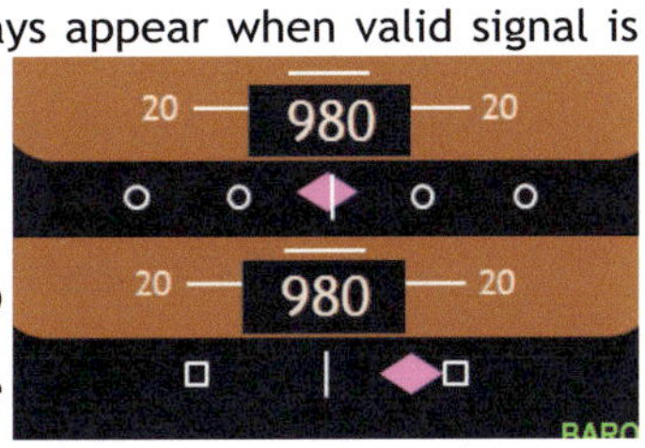

Localizer position indicator
Empty magenta initially
Filled magenta when within 2½ dots from center
Expanded scale--Occurs with FD or A/P LOC engaged ***and***
◆Deviation slightly more than ½ dot ***and***
◆Track < 5° of MCP course
Square marker is ½ dot deviation
With tip of diamond touching either square marker ≈1/3 dot deviation
Deviation to edge of expanded scale causes
Solid diamond to be parked at edge of expanded scale ***and***
To be unfilled if deviation ≥ 2.4 dots on standard scale
Reversion from expanded to normal scale occurs when out of LOC mode ***and***
◆Groundspeed less than 30 knots ***or***
◆Radio altitude greater than 200'

Display Control

EFIS Control Panel

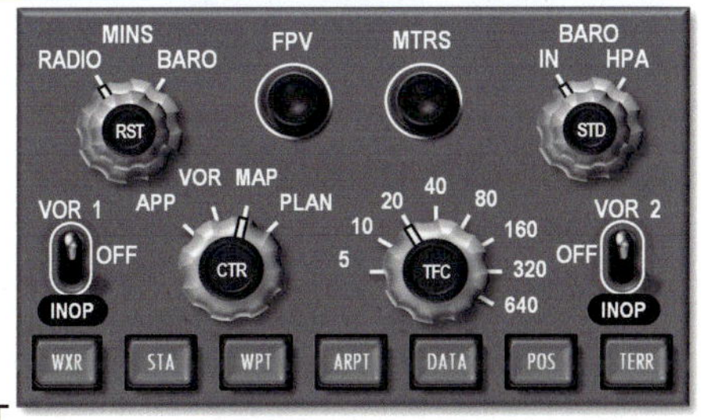

Used for control of CA and F/O outboard and inboard DUs
Either can control all DUs (see Display Source Switch, below)
Switchology is self-explanatory
See FLIGHT INSTRUMENTS, DISPLAYS 30.1

DISPLAYS SOURCE Selector

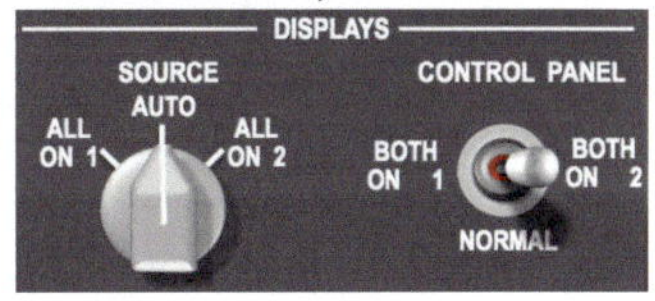

AUTO—Display Electronic Unit (DEU) 1 controls CA outboard & inboard and upper DUs
DEU 2 controls the F/O outboard, inboard, and lower display units
When a DEU fails, remaining DEU controls all DUs
ALL ON 1 or **ALL ON 2**—manually switches to a single DEU for all six display units
Note: *Used on the ground for maintenance purposes*

CONTROL PANEL Select Switch

NORMAL—L EFIS Control Panel controls the CA displays; R EFIS Control Panel controls F/O displays
BOTH ON 1; **BOTH ON 2**—Manually switches CA & F/O displays to a single EFIS Control Panel

HUD Annunciator Panel

F/O Forward Panel

Aids in monitoring of specific landing parameters

Annunciations

NO AIII—Loss of AIII capability
EXCESS DEV—Excessive deviation detected for AII or AIII
TAIL STRIKE—Tail strike event predicted during flare
SYS FAIL—Critical system fault detected
NO FLARE—Flare guidance capability lost
AIII—AIII mode active; required systems and equipment valid
FLARE—Flare guidance is active

3DM & Subsequent Annunciations

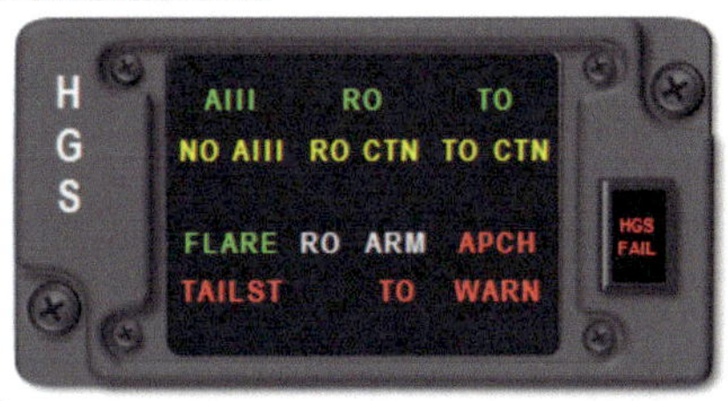

Same as above, except:
APCH WARN—System or approach conditions out of tolerance
RO ARM—System capable of providing ground roll guidance
Displayed prior to touchdown during an AIII approach
RO CTN—Loss of rollout guidance capability
RO—Rollout guidance is active
Other lights—Test, but not used: **TO** **WARN** **TO** **TO CTN**

HGS FAIL Light

Indicates HGS failure
Pushing light causes it to extinguish ***and***
◆Prevents illumination again until failure is cleared and another failure occurs
◆Resets system for another failure

Flight Management, Navigation

Overview

Functional Concept
Each independent of the others
Each can be used singly or in combination
FMS refers to integration of separate components

Components
Flight Management Computer System (FMCS)
Autopilot / Flight Director System (AFDS)
Autothrottle (A/T)
Inertial Reference Systems (IRS)
Global Positioning System (GPS)

Flight Deck Controls
AFDS MCP
Control Display Units (CDU's)
Electronic Flight Instrument System (EFIS) Control Panels,
FMC Source Selector Switch

Primary displays
CDUs
Outboard, Inboard and Upper display units

Data Sources

Global Positioning System (GPS)

Receivers
2 Independent receivers
Operation is automatic

Global Positioning System (GPS) Light MC IRS

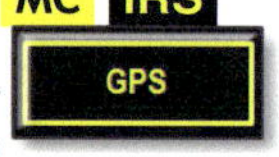

Failure of <u>both</u> GPS sensor units ***or***
Failure of a <u>single</u> GPS sensor unit when either System Annunciator Panel is pressed to initiate a recall
With single GPS sensor failure, light extinguishes when system recall is reset

Display Selector (DSPL SEL)

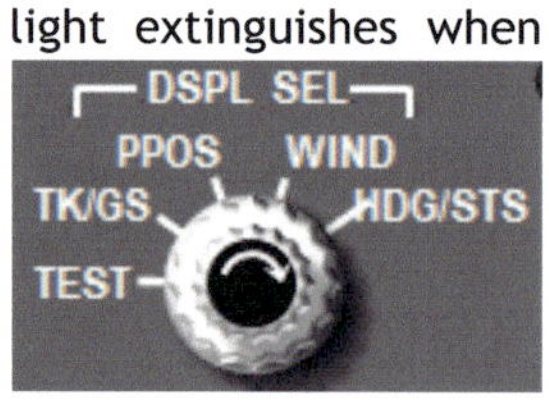

TEST spring-loaded to **TK/GS**
Tests all lights in data displays and mode selector momentarily
Ten second self-test follows
Use only during alignment
TK / GS, **PPOS**—Self explanatory
HDG/STS—L window displays present true heading
R window displays applicable maintenance status codes
During alignment, R window displays minutes to completion
For alignments over 15 minutes, displays 15 until time remaining reaches 14 minutes
Display then counts down in one minute intervals

Keyboard
Alpha key
Keyboard controls data displays when **N**, **S**, **E**, **W** (latitude / longitude) or **H** (heading) are pressed
Pressing an alpha key arms keyboard for numeric entry
Numeric keys
ALIGN Light on—manual present position entry possible
Either mode selector in **ATT**—Permits magnetic heading entry

ON DC Light.......................... MC IRS

AC power abnormal—related IRS is operating on DC power from the switched Hot Battery Bus
On ground—ground-call horn (NW well) sounds
Momentary illumination normal during alignment self-test
If right IRS **ON DC**--power is removed after 5 minutes

DC FAIL Light........................ MC IRS

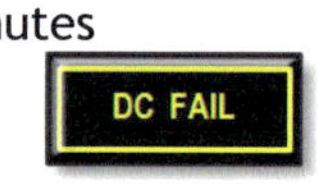

DC power for the related IRS is not normal
If the other lights are extinguished, IRS is operating normally on AC power

Navigation Source Transfer Switches
Inertial Reference System (IRS)
VHF NAV Transfer Switch
BOTH ON L or **R**—Switches appropriate function source to respective side
NORMAL—Attitude and heading source is default IRS

Inertial Reference System (IRS)

Air Data Inertial Reference Units (ADIRU)—Each has an air data system and an IRS system
Air Data system covered in Flight instruments chapter
IRS Systems—2 independent units
Each has 3 sets of ring laser gyros & accelerometers
PFDs display IRS attitude and heading data
Data independent of external NAV aids

Controls & Displays
Inertial System Display Unit (ISDU)
IRS Mode Select Unit (MSU)
IRS Transfer Switch

Alignment
Must be initialized with airplane stationary
Present position must be entered before navigation mode can be entered
Time varies from 5 to 17 minutes depending on latitude
CDU messages
VERIFY POSITION Message—displayed when lat / long not within 4NM of origin airport
ENTER IRS POSITION—Entered lat / long does not pass IRS internal comparison test
Latitude limits
0 to 78° 15' N or S—normal alignment using MSU switch **OFF** to **NAV**
0 to 82° N or S—Magnetic variation for this range stored in IRS memory
Warning Lights

ALIGN Light
<u>Steady</u>—Related IRS I **ALIGN** mode, initial **ATT** mode, or shutdown cycle
<u>Flashing</u>—Alignment cannot be completed due to:
Significant difference between previous and entered positions ***or***
Present position—Unreasonable or none entered

Extinguished
IRS not in **ALIGN** mode
Mode selector **NAV**—alignment complete, all IRS information available
Mode selector **ATT**
Attitude information available
Heading information available after entry of initial magnetic heading

FAULT—system fault detected in related **ATT** or **NAV** mode....................... **MC** **IRS**

Power Sources	Primary	Backup
L IRS	AC Standby Bus	DC from Switched Hot Battery Bus*
R IRS	AC Transfer Bus 2	
Initial Align	Battery Bus (Required)	

*Causes **ON DC** light (and nose gear horn if on ground)

Loss of Alignment
Power—loss of both AC and DC power causes loss
If lost in-flight—
PP and G/S lost for remainder of flight
Attitude can be regained by selecting **ATT** with 30 seconds straight & level flight
Heading—can be manually entered from STBY compass
Must be periodically updated
Entry on **POS INIT** page of FMC/CDU or in ISDU
IRS Mode Selector **OFF**
Remains powered for 30 seconds
ALIGN light ON until completely shut down

IRS Loss Consequences

If either IRS not in **NAV**, (3AA-3DL) the following not available:
- ♦Autopilot
- ♦Autobrakes
- ♦HUD Modes AII and AIII not available

Therefore, no CAT II or III ....QRH FLT-MGT, NAV 11.8

Either IRS not in NAV, (3DM & subsequent):
- ♦IRS ATT and/or NAV mode(s) <u>may</u> be inoperative
- ♦Autopilot should be selected to side with operational IRS
- ♦Autopilot use authorized ***<u>except</u>*** during approach
- ♦If A/P is engaged when failure occurs, yaw damper disconnects after approximately 1 minute; can only be reengaged after A/P is disengaged ....QRH FLT-MGT, NAV 11.12 through 11.16

Right IRS **ON DC**--power removed after 5 minutes (see above)

Radio Navigation

General Features

NavAid Identifier Decoding

Converts Morse Code signal to alpha characters
Not compared to NAV database
Can incorrectly translate characters due to signal variances
Identifier *must* be verified
Audio monitoring not required

Position Updates—FMC autotunes for position updating

Compares two DMEs or a collocated VOR/DME
Identifiers for tuned stations are on **FMC NAV STATUS** page 1/2
Radio position displayed on **POS REF** page 2/3
Specific DME station tuning for position updates can be inhibited on **NAV OPTIONS** page 2/2

Components

DME—Two frequency scanning systems installed

ILS—Two receivers
Manually tuned on VHF Navigation Control Panels
Can be displayed on standby attitude indicator
LOC updating of position occurs through FMC after manual tuning
Standby Attitude Indicator and ISFD ILS information is from No. 1 ILS receiver

Marker Beacon

Displays outer (OM), middle (MM) and inner marker (IM)
Upper right corner of both PFDs
No tuning required

VOR

Must be manually tuned on NAV control panel for display on respective pilot DUs
L and R VOR bearings displayed on DUs when
Valid in range VOR tuned ***<u>and</u>***
Respective EFIS has VOR position selected

Transponder• —(*Most functions similar to standard panels on other airplanes—notes on 737-unique items only*)

Altitude (ALT) **SOURCE**
Selector—enables altitude reporting from respective air data computer (1 or 2)
XPNDR—transponder operates with altitude reporting
Transmissions automatically enabled when air/ground system indicates air mode
Some transponder functions enabled on ground for use at airports with ground scanning position indicating systems
TCAS system described in WARNING chapter

ADF—Not installed

Flight Management Computers (FMC)

General

Certification—"Sole Source" Navigation System

Can operate outside radio NavAid coverage
Second FMC serves as backup

FMC Source Selection

BOTH ON L or **R**—Selects respective FMC for all FMC operations
Right map will annunciate "**FMC L**" and vice versa
NORMAL—L FMC controls CDUs & provides autothrottle input
R FMC operates in synchronization with left FMC
Maps display composite information from both FMCs

<u>Note</u>: *Moving Source Select Switch will causes LNAV / VNAV disengagement*

FMC Alert Light—Message exists for both CDUs, ***<u>or</u>***

TEST Switch is in position **1** or **2**
Press—Both pilots' **FMC** alert lights extinguish

Control & Input

Position updating

Radio autotuning and updating managed by primary FMC
When not available, FMC uses IRS position reference
When external updating not available, position accuracy may be less than required (RNP)

Navigation Performance

FMC Position—represents FMC's estimate of actual position
Current position accuracy derived mathematically from
♦IRS ♦GPS ♦Radio NavAids

<u>Actual Navigation Performance</u> (ANP)—FMC's estimate of quality of position determination
Shown on **POS SHIFT** page 3/3 and **POS REF** page 2/3
Estimated maximum position error with 95% probability
A/C will be less than that distance from displayed points 95% of the time
Lower ANP indicates higher FMC confidence in estimate

<u>Required Navigation Performance</u> (RNP)
FMC supplies default RNP for flight phases

Environment	Default RNP	Time to Alert
Oceanic	12.0 NM	60 sec.
En route	2.0 NM	30 sec.
Terminal	1.0 NM	10 sec.
Approach	0.3 NM	10 sec

UNABLE REQD NAV PERF—RNP message—exceedance indication
RNP available on pages **POS SHIFT, RNP PROGRESS** 4/4 and **RTE LEGS**

Control Display Units (CDUs)

Simultaneous entries from L and R CDUs "should be avoided"
Each CDU loads data to FMC (shared data, independent display)

Fix Entries

Selected entries below; See FLIGHT MANAGEMENT, NAVIGATION pages 40.12 and following
Runway-Related fixes
- Several identifiers available
- Some common ones include
  - Runway threshold—**RWXXX** **RW17L, RW04**
  - Outer Marker—**OMXXX** **OM17L**
  - Touchdown point inboard of RW—**TDXXX**............ **TD17L**

Manually entered latitude / Longitude
- N47° W008° **N47W008**
- N47° 15.4' W008° 3.4' **N4715.4W00803.4**
- These are displayed as **WPT01**, **WPT02**, etc.

Place-Bearing / Distance **SEA330/10**
Place-Bearing / Place-Bearing **SEA330/OLM020**
- Above two would be displayed as **SEA01**, **SEA02**, etc.

Along track before fix........................... **TTT/-10, BOYDS/-5**
Along track after fix **TTT/10, BOYDS/5**

To fly inbound to a fix via a radial
- ◆Line select or type fix to place it in the scratchpad
- ◆Line select CDU 1L OSB (option select button)
- ◆Type intercept course (reciprocal of radial) if assigned to fly inbound via XXX radial
- ◆Line select CDU 6R OSB
- ◆Verify and execute

To fly outbound via a radial
- ◆Line select or type fix to place it in the scratchpad
- ◆Line select CDU 1L OSB
- ◆Enter a point well beyond what is expected to be needed as a "fly-to" point on the radial, i.e. **BOS090/99**
- ◆Line select typed 99 mile point to CDU 2L OSB
- ◆Line select 2L, then 1L to place it first on the LEGS page
- ◆Select the desired course that is already in via 6R OSB
  - *Note that the course in 6R will be magnetically corrected for that geographic point*
  - *That course may come up as much as 4 degrees from the cleared radial, in the example 090R*
  - *The number in 6R will be correct for the station magnetic deviation, however, and should be used*

Newer airplanes—brightness control is a rocker switch

FMC Databases (2)

Performance database

Eliminates need for performance manual reference during normal ops by including:
- ◆Airplane drag and engine characteristics
- ◆Maximum and optimum altitudes
- ◆Maximum and minimum speeds

Maintenance personnel can refine database by entering correction factors for drag and fuel flow

Navigation database

Includes most information normally contained on navigation charts including:
- ◆The location of VHF navigation aids
- ◆Waypoints
- ◆Airports
- ◆Runways
- ◆Other airline selected information, such as SIDs, STARs, approaches, and company routes

Permanent database—retained in computer memory storage
Supplemental database—Data entered manually, remains in storage until deleted (pilot action)
Temporary database—Data entered manually, erased at flight completion
Limits of storage—
- Supplementary and Temporary databases share capacity for 40 navaids and six airports
- 20 waypoint slots reserved in temporary database for entries on that leg
- Remaining 20 slots shared by Supplementary and Temporary databases
- First-come, first-stored basis
- Includes those created on RTE or LEGS pages
- When storage capacity full, unneeded entries must be manually deleted before further waypoints are entered

Two sets of navigation data, each valid for 28 days
- Update time is 0900Z of first day of new database date range
- Coordinated with Jeppeson revision cycles

Engine Out SIDs

Allows display of runway-specific E/O procedures
- E/O SIDS (if any exist) initially display below normal SIDs for airport
- When runway selected, other runways' E/O SIDs are removed from options

If engine fails after takeoff, FMC puts E/O SID into route as a modification if all of the following are true:
- ◆An engine-out SID exists for the ACTIVE departure runway
- ◆An engine-out SID is not already selected for the active route
- ◆The flaps are not up and have not been up since the takeoff was started
- ◆Flight phase is takeoff or climb
- ◆The airspeed is greater than 80 knots (airborne)

Scratchpad message **ENG OUT SID MOD** displays, confirming E/O SID exists for selected runway and was automatically inserted into flight plan — Engine-Out SID (EO SID)
....................on Departure AOM 7.2.2; 6.5.6, p. 6-44
................on Missed Approach AOM 11.2.2, p. 11-11

FMC Failure

Single FMC indications

FMC Alert Light and FMC message relating to specific failure mode
May be caused by loss of data such as TAT
Loss may be temporary
SINGLE FMC OPERATION in scratch pad of both CDUs
"**VTK**" display on failed side NAV display
Switching to **BOTH ON** [non-failed side] may restore operation

Dual FMC Failure

FMC Alert Light and FMC message on both CDUs
Menu page appears on both CDUs
"**VTK**" display on both NAV displays
ASM Flight Management, Navigation 11.32.6

Fuel System

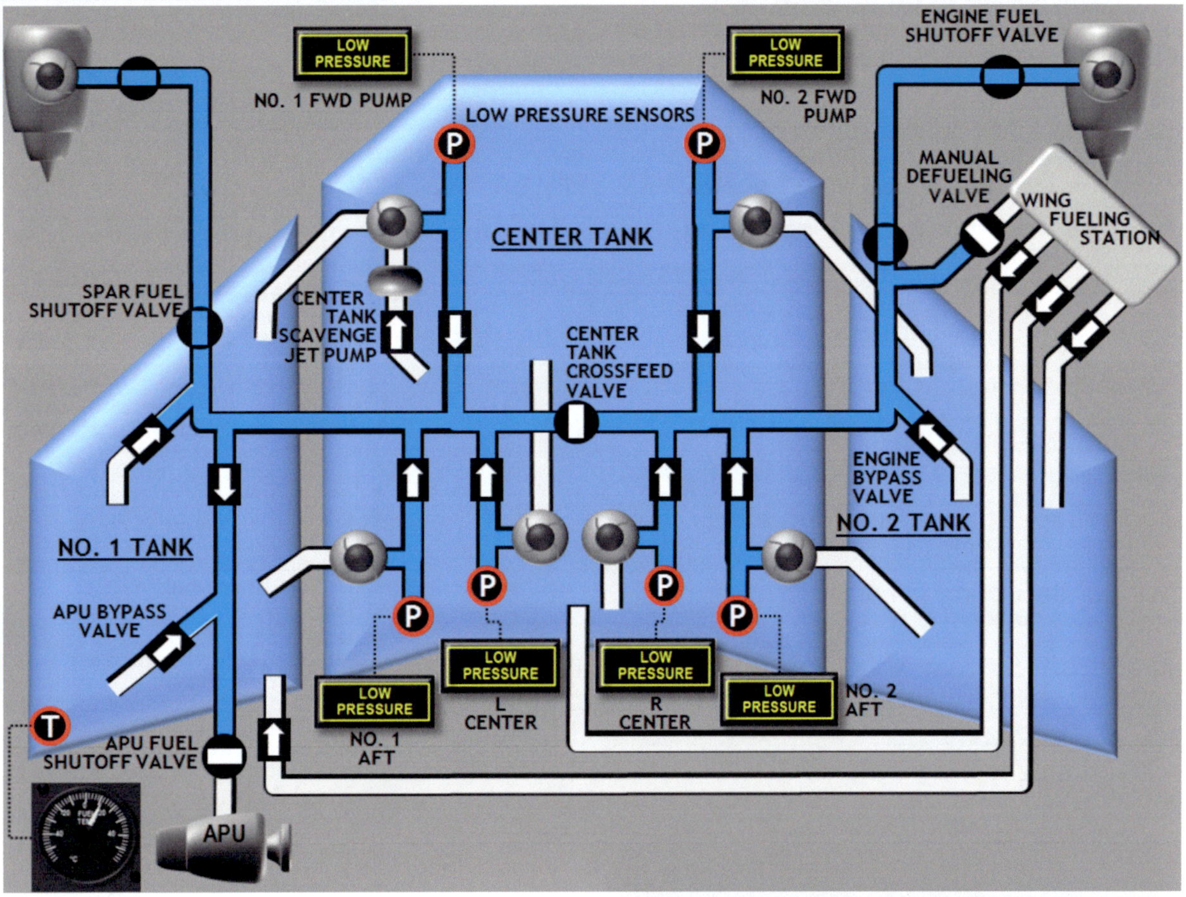

General

Introduction

Supplies pressurized fuel to the engines & APU

Fuel contained in three tanks located within the wings and wing center section

Fuel Feed Normal Order

- Center tank until empty
- Respective main tanks
- Check valves located throughout system to ensure
  - Proper direction of fuel flow ***and***
  - Prevent fuel transfer between tanks

Fuel Feeding

Fuel Pumps

Two AC-powered pumps in each tank

Cooled & lubricated by fuel

Center tank pumps—higher pressure than main tank pumps

- Ensures center tank fuel used before main tank fuel when all fuel pumps operating

Individual pressure sensors *(3DM & subsequent)*

- Monitor each pump's pressure output
- Each center tank pump automatically shuts off (after a short delay) when low pressure detected

<u>Note</u>: *Fuel pump* **LOW PRESSURE** *Lights may flicker when tank quantity is low and the airplane is in a climb, descent, or on the ground with a nose-down attitude*

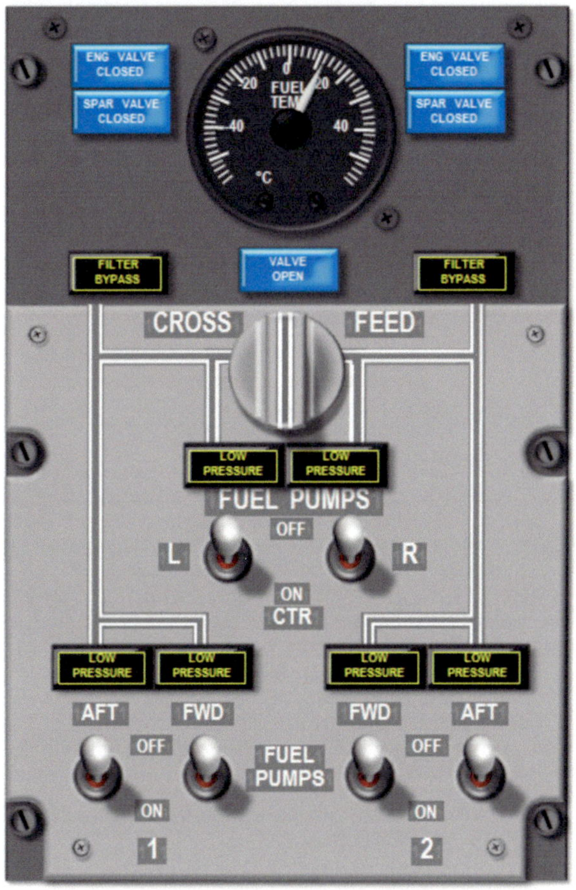

Suction Feed

- With low main tank pressure engines can suction feed
  - Suction feed line bypasses fuel pumps
- Dissolved air
  - In climbs, dissolved air is released from the fuel in the tank due to the decrease in air pressure
  - May collect in the suction feed line and restrict fuel flow
  - High altitude—thrust deterioration or engine flameout may occur
  - Dissolved air will eventually deplete during cruise
  - Depletion time is dependent on: ◆ Altitude ◆ Fuel temperature ◆ Type of fuel
  - After dissolved air is depleted, the engine may be able to suction feed at cruise power
- Main Tank Bypass Valves also used for suction defueling

Fuel Crossfeed

Crossfeed Valve powered by DC from Battery Bus

Used to correct imbalances

Fuel Shutoff Valves

Spar Fuel Shutoff Valves are located at the engine-mounting wing stations

- Power—DC from Hot Battery Bus

Engine Fuel Shutoff Valves

- Fuel actuated, solenoid controlled valves
- Powered from Battery Bus

Closing both valves accomplished by:

- Pulling Fire Switches
- Engine Start Levers to **CUTOFF**

Center Tank Fuel Scavenge Jet Pump

Removes remaining fuel left in center tank

Activates when main tank No. 1
FWD switch is ON *and* quantity is about one half

Once process begins, it continues for the remainder of the flight

Pump is an ejector that uses fuel pressure moving through a spray nozzle to draw fuel out of center tank—no moving parts

APU Fuel Feed

With AC fuel pumps operating, APU fuel is supplied from the left side of the fuel manifold

With AC fuel pumps *not* operating, fuel is suction fed from Main Tank No. 1

Fuel Subsystems

Fuel Temperature

FUEL TEMP Indicator located on the Fuel Control Panel

Sensor in main tank No. 1

System uses AC electrical power

Fueling, Defueling & Ground Transfer

Allows rapid fueling and defueling
Single-point pressure fueling station in the right wing
Also used for the ground transfer of fuel between tanks

Manual defueling valve
Located outboard of engine No. 2
Connects the engine feed system and fueling station

A Shutoff System is used during fueling to automatically close the fueling valve in each fuel tank when full

Fuel Tank Location & Capacities (Usable Fuel)

Tank	Gallons	Pounds*
No. 1	1288	8,600
No. 2	1288	8,600
Center	4299	28,800
Total	6875	46,000

Notes: *These figures represent approximate amounts of usable fuel. The appropriate weight and balance control and loading manual gives exact figures for all conditions.*

**Usable fuel at level attitude, fuel density =6.7 lbs. per U.S. Gallon*

Nitrogen Generation System (NGS)

(3DM & subsequent)

Converts bleed air to nitrogen-enriched air (NEA)

Functions during all phases of flight

NEA delivered to center tank to reduce flammability

Operation is transparent
Automatically starts after take-off
Runs continuously through all phases of flight and during taxi for "a short period of time"
Shuts down after "a specified period of time" *OR*
When bleed pressure is no longer available
Also automatically shuts down during the following non-normal conditions:
◆In Flight: Either engine not running
◆Any time: Left pack overheat
Fire or smoke detected in cargo or main deck areas
◆Ground only: Center tank refueling valve open
Aircraft on the ground, not in test mode

Can be dispatched with INOP NGS—Fuel tanks primarily protected by precluding ignition sources

The NGS operability indicator in right main wheel well adjacent to APU fire control panel

NGS Indicator Lights
OPERATIONAL Light (green)
NGS is fully operational
DEGRADED Light (blue)
NGS is operational, but condition degraded
INOPERATIVE Light (amber)
NGS is inoperative
No lights illuminated—also indicates inoperative

MAX Fuel Flow Alert

Possible fuel leak condition
Displays on inboard display unit

Display Indications

Normal Fuel Display

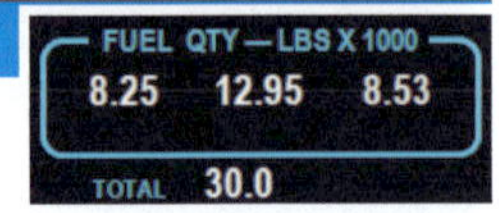

Individual tank and **total fuel** displayed

Electrical power—Standby AC power

Abnormal Fuel System Displays

Fuel LOW Alert

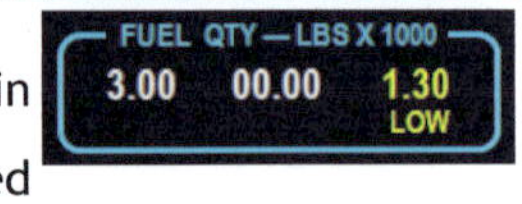

◆Fuel quantity less than 2000 lbs in related main tank
◆Displayed until quantity is increased to 2500 lbs
The fuel quantity on tank(s) with low fuel quantity—AMBER

Fuel Configuration (CONFIG) Alert

◆Either engine running; ***and***
◆Center fuel tank quantity greater than 1600 lbs; ***and***
◆◆Both center fuel tank pumps producing low or no pressure; ***or***
◆◆Both center fuel tank pump switches positioned **OFF**
The quantity digits on center tank indicator turn AMBER
Display remains until -
◆Both engines not running
◆Center fuel tank quantity less than 800 lbs; ***and***
◆One center fuel tank pump producing high pressure; ***or***
◆One center fuel tank pump switch **ON**
The quantity digits on the center tank fuel quantity indicator return to normal

Fuel Imbalance (IMBAL) Alert

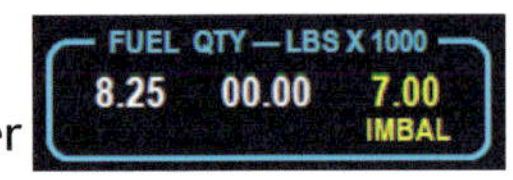

◆Main tanks differ by over 1000 lbs
◆Displayed below main tank with lower fuel quantity
◆Inhibited when airplane is on ground
◆Inhibited by fuel LOW indication when both indications exist
◆Displayed until imbalance is reduced to 200 lbs
Quantity display for tank with less fuel turns amber

MAX Fuel Alerts
FUEL DISAGREE—Sensed vs. calculated quantity disagreement
USING RSV FUEL and **INSUFFICIENT FUEL** alerts (self explanatory)
All three display on FMC and flight display panels

Operational Notes

APU Fuel Use

Extended APU operation on the ground
Position an AC operated fuel pump **ON**
If center tank fuel use desired, use left center tank fuel pump
Prevents a fuel imbalance before takeoff
Maintenance or cockpit crewmember must monitor
With AC electrical power available, use at least one boost pump to supply pressurized fuel to APU
(Extends service life of APU fuel control unit)
See restrictions on center tank fuel use, below

Center Tank Fuel

Do not use:
Ground—with center quantity < 1000 lbs
With flight deck unattended ....AOM 61f.1.2, p. 61f-4; 145.4.1

Pump Selection
Center tank pump inlets staggered
Right pump mounted at rear spar
Left inlet is forward in tank
With small amounts of fuel in center tank, use R pump with nose up pitch attitude and vice versa

During climb
Set the affected center tank fuel pump switch to OFF when center tank fuel pump **LOW PRESSURE** light illuminates.
Set both center tank fuel pump switches to OFF when a center tank fuel pump **LOW PRESSURE** light illuminates if the center tank is empty.

When established in a level flight attitude
If the center tank contains usable fuel and a center tank fuel pump switch(es) is **OFF**
Set the center tank fuel pump switch(es) to **ON** again.
Set affected center tank fuel pump switch to **OFF** when a center tank fuel pump **LOW PRESSURE** light illuminates.
Set both center tank fuel pump switches to **OFF** when a center tank fuel pump **LOW PRESSURE** light illuminates if the center tank is empty. ....AOM 36f.1.2, p. 36f-3

Hydraulics

General

Hydraulic System Power & Backup

Powered by both systems
- Flight controls
  - Ailerons, Rudder, Elevators & Elevator Feel
  - Spoilers—(Flight)—Two On Each Wing On Each System
- Either A or B can power all flight controls
- No decrease in airplane controllability

Backed up by another system

A System		B System
Autopilot A	←→	Autopilot B
Flight Spoilers	←→	Divided between A & B
Landing Gear	←	LG Transfer Unit
NW Steering	←	Alternate NW Steering
Brakes—Alternate	→	Brakes—Normal
#1 Reverser	←STBY→	#2 Reverser
		Autoslats, LE Flaps & Slats —STBY System & PTU
		TE Flaps—Electric Motor

Not backed up by other System

A System	B System
Ground Spoilers	Yaw Damper*

**No hydraulic backup, but standby yaw damper provides functional backup*

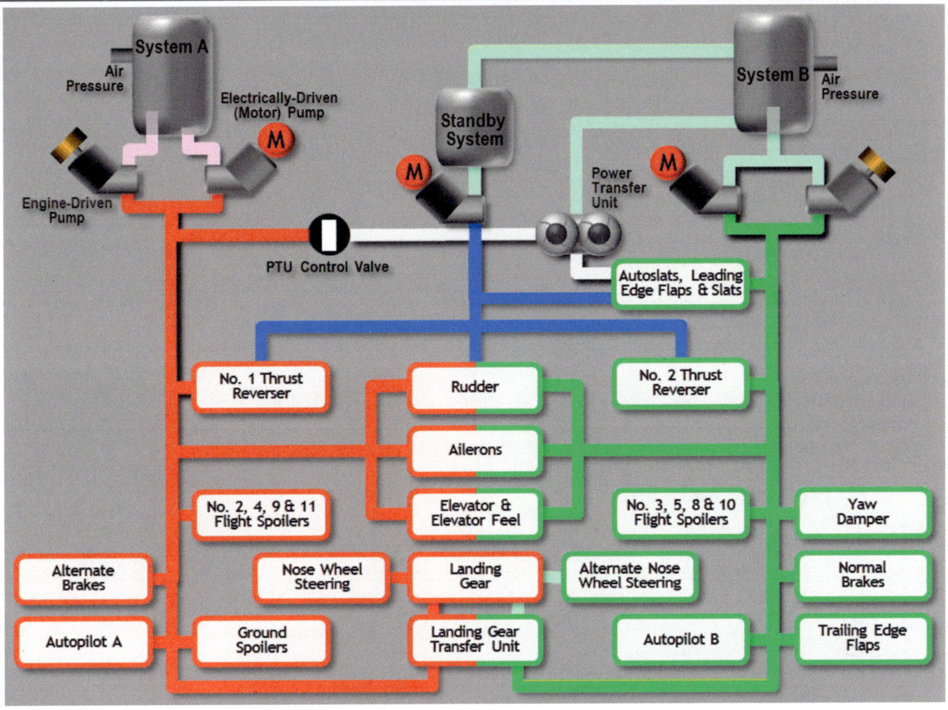

See operating manual for MAX differences

Three Principle Backup Systems

Landing Gear Transfer Unit—Allows raising of gear if #1 engine RPM insufficient to power System A

Power Transfer Unit—In case of low System B engine driven pump output, ensures power for normal rate LE device operation
- ◆Normal rate deployment provides stall protection
- ◆Normal rate retraction prevents excessive drag during acceleration scenarios involving engine 2 failure

Standby System—Allows backup of key flight controls with loss of System A and/or B

Each of these are covered in more detail below

Reservoirs

One for each system, A B & Standby

Located in the main wheel well area

System A and B Reservoirs pressurized by bleed air

Standby Reservoir is connected to the System B Reservoir for pressurization and servicing

Pressurization ensures positive fluid flow to hydraulic pumps

Standpipe Fluid Reservations
- System B—Standpipe fluid for PTU
- System A—flight control protection

Indications

Quantity & pressure displayed with **MFD SYS** button
- Second press blanks indications

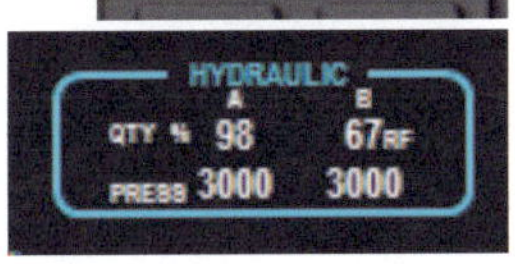

Normal pressure—3,000 psi

Maximum pressure—3,500 psi

RF indication—Refill required
- Quantity below 76%
- Valid only on ground with both engines shutdown *or*
  - After landing with flaps up during taxi in

System Pumps

2 Engine Driven Pumps
- One in each engine accessory drive ....Engine 1—System A ....Engine 2—System B
- Provide approximately 6 times fluid volume of electric pumps
- Turning Pumps **OFF**
  - Isolates fluid from system components
  - Pump still rotates mechanically with accessory drive

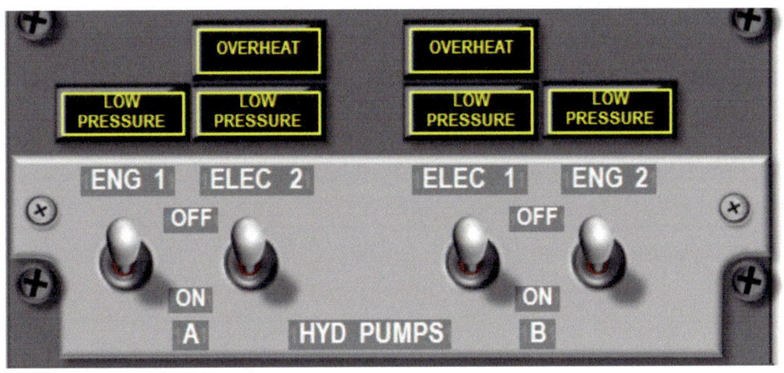

2 Electrical Pumps
- Electric 2—System A........................Electric 1—System B
- Powered by transfer bus on opposite side from engine
- Overheat of electric pump
  - Related **OVERHEAT** Light illuminates
  - Power is removed
- **LOW PRESSURE** Light illuminates

Loss of System A Engine Driven Pump

Resulting Indications
- Low Pressure Light........................ MC HYD

With heavy System A demand, additionally:
- Electric pump volume lower—Intermittent Electric 2 **LOW PRESSURE** Light
- Master Caution and Annunciator ...... MC FLT CONT and HYD

Cooling & Lubrication

Hydraulic fluid is used

Passes through heat exchangers
- System A heat exchanger in Main Fuel Tank No. 1
- System B heat exchanger in Main Fuel Tank No. 2

Minimum main tank fuel for cooling—1675 Lbs

Overheat Light—cooling fluid or pump itself has overheated and pump has shut down.................... MC HYD

Pressure Indications

Pressure switches

In engine-driven and electric motor-driven pump output lines

Send signals to illuminate related **LOW PRESSURE** Light if pump output pressure is low

Pressure transmitter—Sends combined pressure of the engine-driven and electric motor-driven pump to related hydraulic system pressure indication

Low Pressure

Check valves Isolate the related pump from the system

Located in each output line

Pressures

Normal Pressures ....3000 psi
Maximum Pressure ....3500 psi
Minimum Pressure (normal operations)................2800 psi

Leaks

System A

Engine-driven Pump and related line leak—standpipe in reservoir prevents loss below 20%

Electric pump or related line and component leak
- Quantity can decrease to zero
- All pressure can be lost

System B

Leak in either pump can decrease indicated quantity to zero

Standpipe fluid retained supplies fluid to both pumps

Remaining fluid is sufficient for Power Transfer Unit operation

Does not affect operation of Standby Hydraulic System

Backup & Redundancy

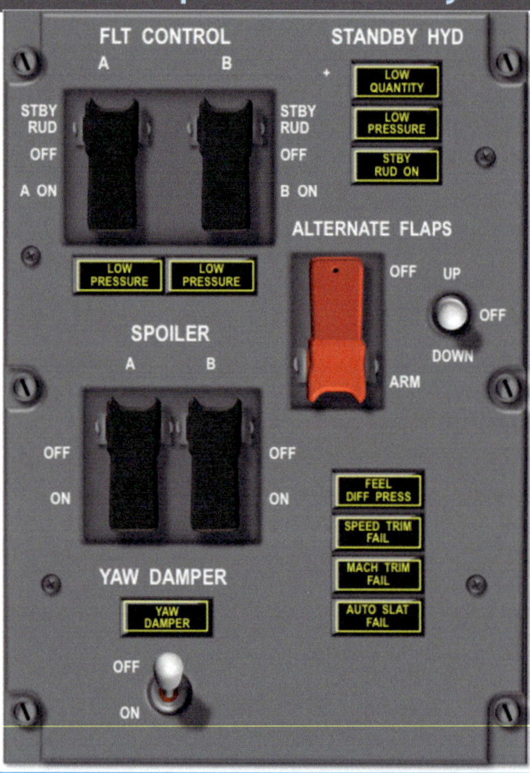

Power Transfer Unit

Purpose—In cases of low output from System B engine driven pump, allows power for LE device operation at normal rates
- ♦Normal rate deployment provides stall protection
- ♦Normal rate retraction prevents excessive drag during acceleration scenarios involving engine 2 failure

Operation

Uses System A Pressure to power a pump

Pump pressurizes System B fluid for flight controls above

Automatic activation when:

- ♦System B Engine-Driven Pump pressure drops below limits
- ♦Airborne
- ♦Flaps are less than 15 but not up

Landing Gear Transfer Unit

Purpose—Supply sufficient volume of hydraulic fluid to raise the landing gear at the normal rate when #1 engine RPM is low (which may lead to low system A engine-driven pump volume output)

Operation—Uses System B engine driven pump output

System A is no longer connected to landing gear actuators

System B operates the gear when activation criteria met

Automatic Activation

A—*Airborne*
N—*No. 1 Engine* RPM drops below a limit value
U—*Up*; LANDING GEAR Lever is positioned UP
L—*Locked*; either main landing gear not up and locked

Standby Hydraulic System

Purpose

Backup if System A and/or B pressure is lost

Powers four key systems (**TRYS**)

♦**T**-*Thrust:*	Thrust Reversers
♦**R**-*Rudder:*	Rudder
♦**Y**-*Yaw Damper:*	Standby Yaw Damper
♦**S**-*Slats:*	Leading Edge Flaps & Slats (extend only)

Operation—Uses a single electric motor-driven pump

Activation—Manual

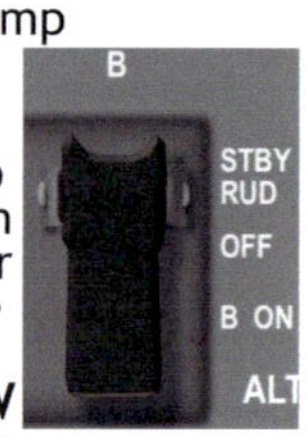

Either **FLT CONTROL** switch to **STBY RUD**
- ♦Activates Standby Electric Motor-Driven Pump
- ♦Shuts off the related hydraulic system pressure to ailerons, elevators and rudder by closing the Flight Control Shutoff Valve
- ♦Opens the Standby Rudder Shutoff Valve
- ♦Deactivates related Flight Control **LOW PRESSURE** Light when Standby Rudder Shutoff Valve opens
- ♦Allows the standby system to power the rudder and thrust reversers
- ♦Illuminates the **STBY RUD ON**, Master Caution, and **FLT CONT** lights

Positioning the **ALTERNATE FLAPS** master switch to **ARM**
- ♦Activates the Standby Electric Motor-Driven Pump
- ♦Closes the Trailing Edge Flap Bypass Valve
- ♦Arms the **ALTERNATE FLAPS** position switch
- ♦Allows the Standby System to power the leading edge flaps and slats and thrust reversers

Activation—Automatic

Initiated when:
- Loss of System A or B, ***and***
- Flaps extended, ***and***
- Airborne or wheel speed > than 60 knots, ***and***
- **FLT CONTROL** Switch **A** or **B** Hydraulic System **ON**

OR:
- The main PCU Force Fight Monitor (FFM) trips

Activation Results: .... **MC** **FLT CONT**

- Activates Standby Electric Motor-driven Pump
- Opens Standby Rudder Shutoff Valve
- Allows Standby System to power rudder and thrust reversers

Standby System Leak

Standby reservoir quantity decreases to zero

LOW QUANTITY light on when reservoir ≈½ empty

Effects on System B
- Continues to operate normally
- Reservoir fluid indication decreases and stabilizes at approximately 72%

Variations in Hydraulic Quantity Indications

Occur when:

Initial pressurization after engine start

Raising or lowering the landing gear or leading edge devices

Cold soaking during long periods of cruise

Normally—Little effect on systems operation

Foaming

Occurs if
- Hydraulic system is not properly pressurized
- At higher altitudes

Symptoms
- Pressure fluctuations
- Blinking of the related **LOW PRESSURE** Lights
- **MC** **HYD** annunciator lights may illuminate momentarily

Landing Gear

Gear System

Overview

Hydraulic Power—System A for extension, retraction, and nose wheel steering

Backups
- Manual landing gear extension system
- Alternate nose wheel steering

Brake System—Hydraulic system B

Alternate Brake System—Hydraulic System A

Antiskid protection available on both normal and alternate brake systems

Autobrakes only available with normal brake system

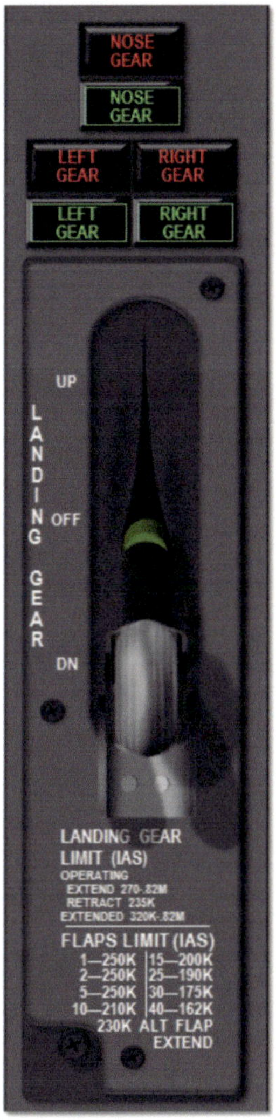

Landing Gear

Extension
- System A releases uplocks
- Gear extends by hydraulic pressure, gravity and air loads
- Held at full extension by over center mechanical and hydraulic locks
- Nose wheel doors remain open when gear is down
- Manual Gear Extension
  - Manual handles on flight deck release uplocks
  - Gear free fall to down and locked position
  - With door open
    - ♦Manual gear extension possible with LANDING GEAR Lever in any position
    - ♦Normal landing gear extension is possible if Hydraulic System A pressure is available
    - ♦Landing gear retraction is disabled
  - Following manual extension, gear may be retracted normally
    - Hydraulic System A pressure must be available
    - Manual extension access door must be closed

Retraction
- Main gear
  - Wheels stopped by brake system as gear retract
  - Held up by hydraulic uplocks
  - Damaged tire detection
    - Loose tread impacts a fitting which signals system to lower that gear
    - Prevents loose tread from damaging wheel well components and lines
- Nose gear
  - Wheels stopped by snubbers in nose wheel well
  - Held up by overcenter lock
- Hydraulic pressure removed when gear handle is moved to **OFF**
- Landing Gear Transfer Unit
  - Uses hyd system B to raise gear
  - Enabled when all of the following exist (**ANUL**):
    - **A**—*Airborne*
    - **N**—*No. 1 Engine* RPM drops below a limit value
    - **U**—*Up*; LANDING GEAR Lever is positioned UP
    - **L**—*Locked*; either main landing gear not up and locked

Landing Gear Indicator Lights (bottom)
- Illuminated (green)—Related gear down & locked
- Extinguished—Landing gear is not down and locked
- LG Warning horn is deactivated with all gear down & locked
- LG is down & locked if either center or overhead panel green light is on

Landing Gear Indicator Lights (top)
- Red—Landing gear is not down and locked *with*
  - ♦Either or both thrust levers retarded to idle, and below 800 feet AGL) ***or***
  - ♦Related landing gear is in disagreement with **LANDING GEAR** Lever position (in transit or unsafe)
- Extinguished—All gear up & locked with Gear Lever **UP** or **OFF**
  - *OR* Landing gear down & locked with Gear Lever **DN**

Landing Gear Indicator Lights

- (Overhead Panel)
- Redundant but separate set of LG indicator circuits and lights
- ON—Related gear down and locked
  - Verified down and locked as long as one green indicator light (center panel or overhead panel) is illuminated
- Extinguished - Landing gear is not down and locked

MAX Differences compared to 737-800
- Panel located in center of forward panel
- Override switch
  - **LOCK OVRD** Releases landing gear lever lock
- Landing Gear Lever
  - Electrical switch that controls hydraulic extend and retract system
  - No mechanical connections to valve system
  - Pressure is automatically removed 10 seconds after all gear up and locked
  - No **OFF** position

Nose wheel steering

Hydraulic Power

- With **NOSE WHEEL STEERING** switch in **NORMAL** —System A
- With switch in **ALT**—System B
  - With a leak downstream of landing gear transfer unit
    - System B fluid loss can occur
    - Sensor closes landing gear transfer valve
    - Alternate NW steering no longer possible
- Available only when airplane is on the ground

Control Limits
- ♦With tiller............................± 78°
- ♦With pedals ....± 7°

Towing Lever
- Towing lever locks out hydraulic pressure from nose wheel steering
- Lockout pin holds lever in place when towing

Brakes

Brake System Power

Normal brakes ....System B

Alternate brakes....................System A
- Automatically supplies pressure if System B is low or fails

Accumulator ....System B pressurizes
- Automatically supplies pressure if normal and alternate system pressure lost
- Trapped hydraulic pressure in accumulator provides
  - Several braking applications ***or***
  - Parking brake application

Hydraulic Brake Pressure (BRAKE PRESS) Indicator
- Indicates brake accumulator pressure:
- Normal pressure - 3000 psi
- Maximum pressure - 3500 psi
- Normal precharge - 1000 psi

Antiskid

Available with:
Both normal and alternate brake systems
Loss of both hydraulic systems

System detects and **releases** brake pressure in event of
- ♦Skid ♦Touchdown
- ♦Locked wheel ♦Hydroplane

Normal system—<u>Each wheel</u>—Individual antiskid protection
Alternate system—<u>wheel pairs</u> have antiskid protection

ANTISKID INOP Light

System fault is detected by Antiskid Monitoring System
Extinguished—System operating normally

Autobrake System

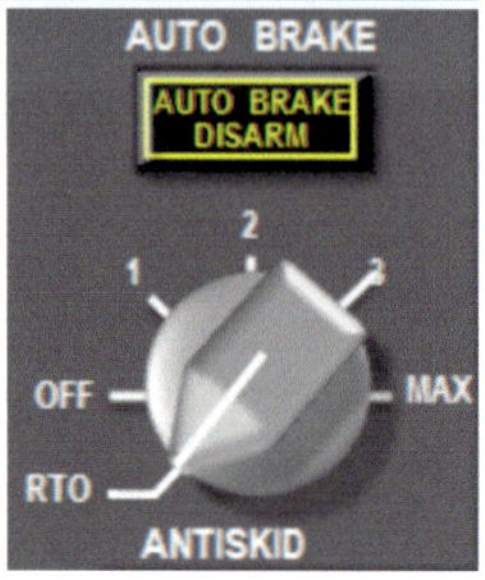

Powered by.......................System B
Available with .....ONLY Normal Brakes
Modes
Rejected Takeoff (RTO)
Maximum deceleration
Can be selected <u>only</u> on ground
Arming required conditions
- Airplane on the ground
- Antiskid and autobrake systems operational
- **AUTO BRAKE** Select Switch positioned to **RTO**
- Wheel speed less than 60 knots
- Thrust Levers positioned to **IDLE**

If **RTO** selected and <u>TO rejected < 90</u> knots
AUTOBRAKE DISARM light does NOT illuminate
RTO Autobrake function remains armed

If **RTO** selected and <u>TO rejected > 90</u> knots
Maximum braking applied ***<u>when</u>***
Thrust Levers retarded to **IDLE**

Disarmed when—both Air/Ground Systems indicate air mode
AUTO BRAKE DISARM Light does not illuminate
AUTO BRAKE Select Switch remains in **RTO**

If NOT disarmed prior to landing
No automatic braking action occurs ***<u>and</u>***
AUTO BRAKE DISARM Light illuminates 2 seconds after touchdown

Landing Modes
Four levels of deceleration selectable.............(1, 2, 3, MAX)
Maximum Autobrake deceleration is less than full pedal braking
Autobrake application begins when:
- ♦Both thrust levers are retarded to **IDLE** ***<u>and</u>***
- ♦Main wheels spin-up

Disarm occurs when:
<u>Any time</u>--**AUTO BRAKE** switch to **OFF**
<u>After braking has started</u>
- ♦**SPEED BRAKE** lever to down detent ***<u>or</u>***
- ♦Advancing thrust lever(s) (except during first 3 seconds after touchdown) ***<u>or</u>***
- ♦Applying manual brakes

AUTO BRAKE DISARM light comes on for all above except **AUTO BRAKE** switch to **OFF**

AUTO BRAKE DISARM Light

Illuminated:
- ♦**SPEED BRAKE** Lever moved to down during RTO or landing
- ♦Manual brakes applied during RTO or landing
- ♦Thrust Lever(s) advanced during RTO or landing, except during first 3 seconds after touchdown
- ♦Landing made with RTO selected
- ♦**RTO** mode selected on ground—Illuminates for 1 to 2 seconds, then extinguishes
- ♦A malfunction exists in automatic braking system

Extinguished:
- ♦**AUTO BRAKE** Select Switch set to **OFF**
- ♦Autobrakes armed

Parking Brake

Power Sources
System A or B if pressurized ***<u>or</u>***
Brake accumulator
Accumulator pressure shown on HYD BRAKE PRESS indicator

Fault in system may cause ANTISKID INOP light

If thrust lever advanced for takeoff with **parking brake set**
All aircraft.................................TO Config warning horn
3DM & Subsequent......................**TAKEOFF CONFIG** lights on

Air / Ground System

6 Sensors—two per landing gear
Numerous systems controlled
Details in chart ....OMV2LANDING GEAR 14.20.6

Warning Systems

Overview

Systems Covered

This section
- ♦Windshear warnings
- ♦Landing gear warnings
- ♦Ground proximity warnings
- ♦Mach / airspeed Warnings
- ♦Takeoff configuration warnings

Other Sections

♦Cabin altitude ....Air Systems
♦Autopilot & autothrottle disconnect..........Automatic Flight
♦Fire..Fire Protection

Color Coding

Immediate attention .... RED
Fire—Engine, Wheel Well, Cargo or APU
Autopilot and autothrottle disconnects
Landing Gear Unsafe

Timely attention.. AMBER

Status.. BLUE
Information about status or configuration change
Some blue lights are bright during transition, dim after

Configuration Confirmation .... GREEN
Gear, Flaps, Speed Brake on forward panel
LE Devices on overhead panel

Master Caution & System Annunciator

CA Side

F/O Side

Master Fire Warning Lights

Remain illuminated as long as condition exists

Reset Feature—Pressing light
Extinguishes both lights
Silences fire warning bell
Resets system for future warnings

Master Caution Lights

Indicate cautionary condition on system outside pilots' normal field of view

Remain on
As long as condition exists ***or***
Until crew resets system

Reset Feature—Pressing light
Extinguishes both lights
Resets system for future cautions
Pressing either System Annunciator light panel recalls all existing fault conditions

Single Fault in Redundant System, or simple fault
Do not illuminate Master Caution lights in some cases
Fault is stored in **MC** system
Pressing System Annunciator Panel recalls the single fault
This is referred to as **MASTER CAUTION** Recall
All 12 lights (6 per side) light when either System Annunciator Panel is pressed

In This Book, MC indicates Master Caution Light in text
System Annunciator panel indications appear as OVERHEAD

System Annunciator Lights

Design
Calls pilot attention to faults and other warning lights outside normal field of view
Areas covered are
Forward And Aft Overhead Panels
Fire Control Panels
Any of covered faults illuminate appropriate System Annunciator Panel ***and*** MC light

Left Side (CA) Glare Shield

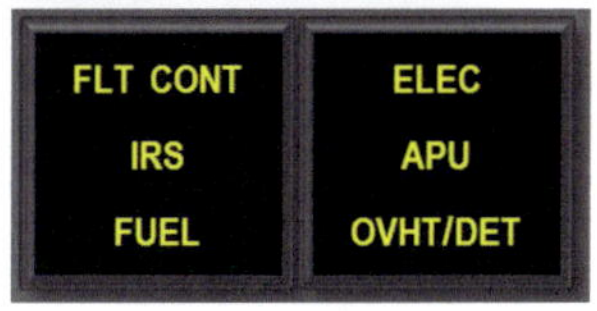

FLT CONT
- ♦ LOW QUANTITY
- ♦ LOW PRESSURE
- ♦ FEEL DIFF PRESS
- ♦ SPEED TRIM FAIL
- ♦ AUTO SLAT FAIL
- ♦ YAW DAMPER
- ♦ MACH TRIM FAIL
- ♦ STBY RUD ON
- ♦ MAX ASSIST ON
- ♦ MAX SPOILERS

IRS
- ♦ FAULT
- ♦ DC FAIL
- ♦ ON DC
- ♦ GPS
- ♦ MAX ILS
- ♦ MAX GLS

FUEL
- ♦ LOW PRESSURE
- ♦ FILTER BYPASS

ELEC
- ♦ DRIVE
- ♦ STANDBY PWR OFF
- ♦ SOURCE OFF
- ♦ TR UNIT
- ♦ TRANSFER BUS OFF
- ♦ BATTERY DISCHARGE
- ♦ ELEC

APU
- ♦ FAULT
- ♦ OVERSPEED
- ♦ LOW OIL PRESSURE
- ♦ MAX DOOR

OVHT/DET
- ♦ ENGINE 1 OVERHEAT
- ♦ APU DET INOP
- ♦ ENGINE 2 OVERHEAT

NO system annunciator for Cargo Fire **DETECTOR FAULT**
Engine Fire Detector Loop **FAULT**

Right Side (FO) Glare Shield

ANTI-ICE
- ♦ WINDOW OVERHEAT
- ♦ PITOT HEAT
- ♦ COWL ANTI-ICE
- ♦ 3AA-3PX PSEU
- ♦ MAX ELT
- ♦ MAX EQUIP SMOKE
- ♦ MAX MAINT

HYD
- ♦ OVERHEAT
- ♦ LOW PRESSURE

DOORS
- ♦ FWD / AFT ENTRY
- ♦ FWD / AFT SERVICE
- ♦ LEFT or RIGHT OVERWING EXIT
- ♦ FWD / AFT CARGO
- ♦ EQUIP

ENG
- ♦ REVERSER
- ♦ ENGINE CONTROL
- ♦ MAX RVSR LIMITED
- ♦ MAX REVERSER AIR/GND
- ♦ EEC ALTN MODE
- ♦ 3AA-3PX REVERSER
- ♦ MAX RVSR COMMAND

OVERHEAD
- ♦ EQUIP COOLING Off
- ♦ FLIGHT RECORDER Off
- ♦ EMER EXIT LIGHTS—Not Armed
- ♦ LAVATORY SMOKE
- ♦ MAX EQUIP SMOKE
- ♦ PASS OXY - ON
- ♦ ELT (3DM & Subsequent)
- ♦ 3AA-3PX PSEU
- ♦ MAX MAINT

AIR COND
- ♦ DUAL BLEED
- ♦ PACK
- ♦ OFF SCHED DESCENT
- ♦ 3AA-3PX BLEED TRIP
- ♦ WING-BODY OVERHEAT ..
- ♦ AUTO FAIL
- ♦ ZONE TEMP
- ♦ MAX BLEED

Configuration Warnings

Takeoff Configuration

Armed when
Airplane on ground ***and***
Either or both thrust levers advanced for takeoff

Intermittent horn sounds for any of the following:
- ♦Trailing edge flaps are not in the flaps takeoff range (*1-25*), ***or***
- ♦Leading edge devices not configured for takeoff, ***or***
- ♦SPEED BRAKE Lever is not in the DOWN position, ***or***
- ♦Spoilers not down with speedbrake lever in DOWN position, ***or***
- ♦Parking brake is set, ***or***
- ♦Stabilizer trim not set in the takeoff rangeWarning indication is cancelled when the configuration error is corrected

Warning indication is cancelled when the configuration error is corrected

TAKEOFF CONFIG Warning Light
On for any of the above

Landing Gear Configuration Warnings

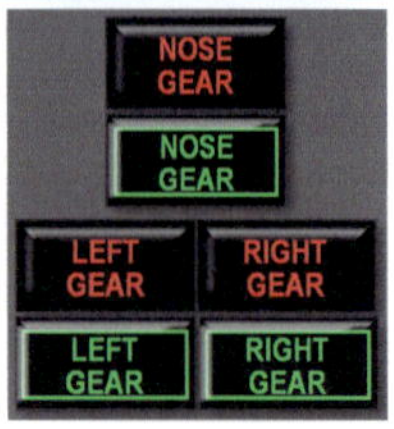

Gear Visual Indications
- Gear Down & Locked.......... **GREEN** Light
- Gear not in agreement with handle .... **RED** Light
- LG not down & locked *AND* < 800' AGL ***and***
  - Either thrust lever retarded to idle .... **RED** Light
- LG Handle **UP** ***or*** **OFF**, ***and*** all gear up & locked ....Lights off

Warning Horn—Steady horn alerts crew
- Flaps up—10° ....May be silenced
  - Below 800'AGL ***and***
  - Low thrust
    - Either thrust lever between idle and 20° lever angle ***or***
    - 1 engine not operating and other less than 34°
- Flaps Extended 15—25° ....Cannot be silenced
  - Any Altitude, with Low Thrust (As above)
- Flaps Extended >25° ....Cannot be silenced
  - Any altitude, any thrust lever setting
- Warning is cancelled when configuration error is corrected

Proximity Switch Electronic Unit (PSEU) 3AA-3PX

Monitors four major systems
- ◆Takeoff configuration ◆Landing gear
- ◆Landing configurations ◆Air/ground sensing
- Also monitors numerous other systems for internal faults
- Indications—**PSEU** Light .... **MC** **OVERHEAD**

Reset—Possible after
- Maintenance BITE check ***or*** repair of cause of the fault

PSEU Inhibited--In flight...
- ◆When the thrust levers are advanced toward takeoff power
- ◆For 30 seconds after landing

Newer Aircraft (3DM & Subsequent)
- Simple faults do not illuminate PSEU
- Pressing System Annunciator Panel RECALL illuminates appropriate system light--may help isolate affected system
- Simple faults
  - May trigger PSEU when system annunciator is pushed for recall
  - In this instance, resetting **MC** system extinguishes PSEU light
- **Note**: *Flight Lock function is discussed in AIRPLANE GENERAL,* *page 24* of this Guide

Maintenance Light (MAINT) 3RA-3VM

- System fault--maintenance action required
- Inhibited ***from*** first engine start
  - ***to*** 30 seconds after landing
- **Note**: *This light is not to be confused with the MAINT menu tab located on the MFD systems (SYS) page. Refer to Chapter 13, Hydraulics.*

Cabin Altitude Warning Light

On if cabin pressure altitude exceeds 10,000'

Activation simultaneous with aural warning intermittent horn for **CABIN ALTITUDE** alert

Tail Skid

Incorporates a crushable honeycomb cartridge

Serviceable when cartridge decal shows both green and yellow
- Green disappears gradually as cartridge is crushed
- When warning decal is all red—cartridge must be replaced

Flight Parameter Warnings

Mach/Airspeed Warning System

Aural Warning—Distinctive "Clacker"
- Indicates VMO/MMO exceedance
- Cannot be silenced except by slowing to acceptable speed

Visual Indication
- Maximum and minimum speeds .... **RED** Warning Band
- Max and min maneuvering speeds....... **AMBER** Caution Band

System Test
- Only possible on ground
- Clacker sounds; no visual warning

Stall Warning System

Indicates approach to stall angle of attack

Nuisance Stall Warning—occurs in specific uncommon situations

Intuitive System .....ASM Warning Systems 15.20.4, p. 15-20-9

Altitude Alerting

Approach to Altitude—900' above or below
- White box around selected altitude
- Momentary tone sounds
- Box disappears at 200' above or below altitude

Deviation Alerting—200' or more off altitude after capture
- Momentary tone
- Current altitude box turns amber and flashes
- Continues until altitude decreases to ≤200' from selected ***or***
  - Altitude increases to >900' from selected ***or***
  - New Altitude selected

Weather Radar

Designed for turbulence avoidance

Detects precipitation velocity < 40NM

Detects windshear events < 1200' AGL and < 3 NM of aircraft
ASM FLT MGT—NAV 11.20.24, Warning Systems S 15.10.13

Bank Angle Alert

GPWS provides Aural Alerts ....*Bank Angle, Bank Angle*

Sounds when bank passes 35°, 40° and 45°

Alert sounds **once**, then silences until bank decreases to ≤ 30°

Approach Callouts

Similar to other aircraft in the AA Fleet

GPWS provides
- Barometric alerts at 1000' and 500' AGL
- Radio Altimeter alerts at 2500', and < 500'

Warning System Test Panel

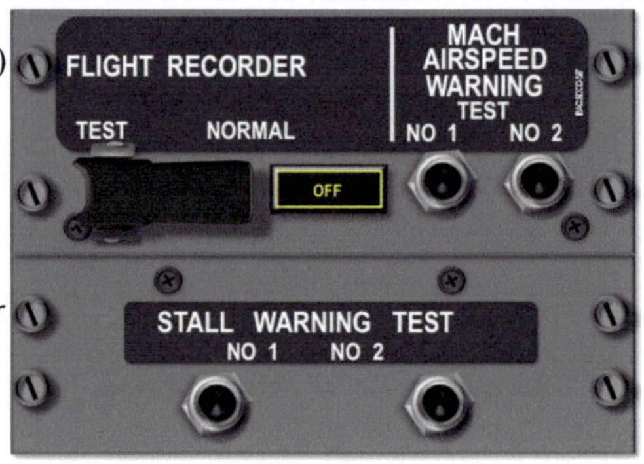

Flight Recorder Test
- **NORMAL** (guarded) position
  - Flight—recorder operates if electrical power is available
  - Ground—Either engine must also be operating
- **TEST**—Powers the flight recorder on the ground.

Mach Airspeed Warning Test
- Respective Mach/Airspeed Warning System clacker sounds
- Inhibited while airborne

Stall Warning Test
- Functions only on ground with AC power available
- Each switch tests respective Stall Management Yaw Damper (SMYD) Computer.
  - No. 1 SMYD computer shakes Captain's control column
  - No. 2 SMYD computer shakes FO's control column
  - Vibrations felt on both columns
- Inhibited airborne

Ground Proximity Warning System (GPWS)

General

Uses Radio Altitude
- Combines with
  - ◆Barometric altitude
  - ◆Airspeed
  - ◆Glide slope deviation
  - ◆Airplane configuration

Alerts provided for
- ♦Excessive descent rate
- ♦Excessive terrain closure rate
- ♦Altitude loss after takeoff or go-around
- ♦Unsafe terrain clearance when not in the landing configuration
- ♦Excessive deviation below an ILS glide slope

Uses internal database of worldwide terrain
Proximate terrain shows on Navigation Display
"Look ahead terrain alerts" warn of alerts based on estimated time to impact
Alerts accompanied by voice alerts (*"PULL UP"*) and PFD displays
Visual displays repeat on HUD

GPWS Inhibit & Control Panel

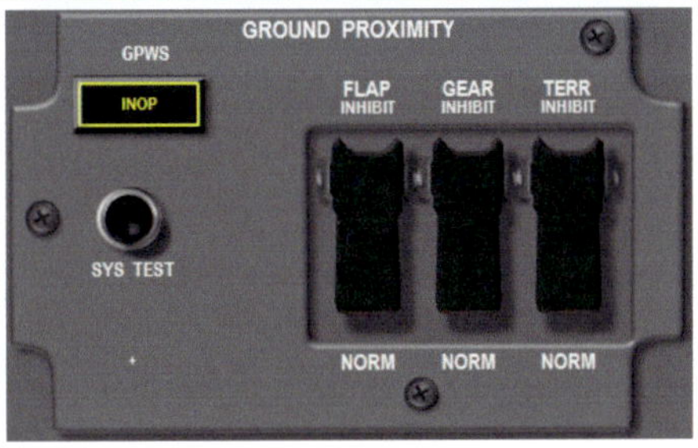

FLAP INHIBIT Switch
Inhibits ground proximity TOO LOW FLAPS alert

GEAR INHIBIT Switch
Inhibits ground proximity TOO LOW GEAR alert

TERR INHIBIT Switch
Inhibits look-ahead terrain alerts and terrain display

GPWS INOP Light—GPWS computer malfunction or power loss
Invalid inputs are being received from any of these: ♦ADIRU
- ♦Stall management computers
- ♦Radio altimeter
- ♦IRS
- ♦EFIS Control Panel
- ♦ILS receiver
- ♦FMC

Below Glide Slope Light

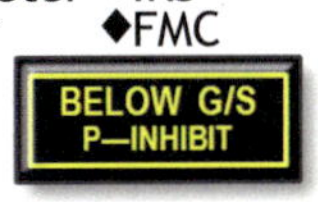

Deviation below glide slope
Accompanied by aural warning, *"GLIDE SLOPE"*
Volume and repetition rate increase as deviation increases
Press—Inhibits ground proximity **GLIDE SLOPE** alert when below 1,000 feet radio altitude

Ground Proximity System Test (SYS TEST) Switch

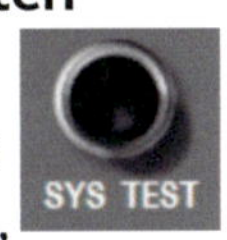

Momentarily on ground:
- ♦**BELOW G/S** and **GPWS INOP** Lights
- ♦**TERR FAIL** and **TERR TEST** on navigation displays
- ♦**PULL UP** and **WINDSHEAR** visual alerts
- ♦*"GLIDE SLOPE," "PULL UP,"* and *"WINDSHEAR"* aurals
- ♦Terrain display test pattern on navigation displays
- ♦*"TERRAIN"* caution aural sounds and **TERRAIN** caution message shows on navigation displays

Above indications are followed by the following aural tests:
- ♦Radio altitude based alerts
- ♦Bank angle alert
- ♦Approach callouts
- ♦Windshear alert
- ♦Look ahead terrain alerts

System test inhibited in flight

GPWS Color Coding

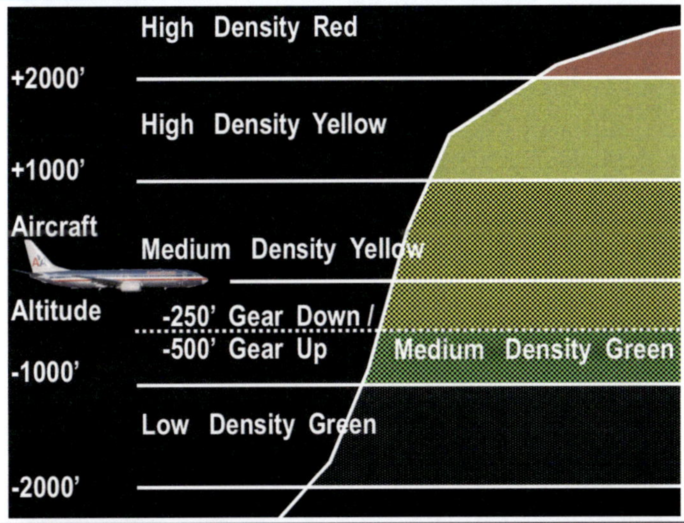

Predictive Wind Shear

Overview

Available during takeoff, approach and landing
GPWS provides warning when A/C in windshear
Weather Radar provides predictive windshear alerts when conditions ahead are detected

Prioritization—Alerts based on relative hazard and required reaction time
Inhibited by
- Actual windshear warning
- Look-ahead terrain alerts
- Radio altitude-based alerts

Activation

Automatic radar scanning begins when
Ground—Thrust Levers set for takeoff
In flight— < 2300' RA

Predictive windshear alerts issued < 1200' AGL

Alerts available ≈ 12 seconds after scanning begins

PWS alerts can be enabled prior to takeoff by turning radar **ON** on ground

Without PWS alert conditions detected—weather radar returns do not display unless EFIS WXR switch pressed

PWS Alert Area

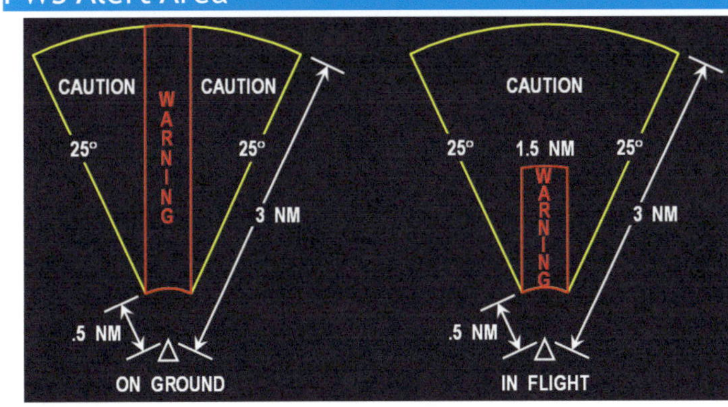

Inhibits on Takeoff and Landing
Cautions .... 80 knots to 400' RA
Warnings .... 100 knots to 50' RA

Traffic Alert and Collision Avoidance System (TCAS)

Controlled through Transponder Panel

Other aircraft detected through transponder
- Non-equipped aircraft not detected
- Independent of ground-based radar

Advisories based on point of closest approach
- Traffic advisory (TA) issued when within approx. 40 seconds
- Resolution advisory (RA) when within approx. 25 seconds

Traffic Symbols and Warnings

Other Traffic ....Open White Diamond
- More than 6 miles away ***or***
- Over 1200' altitude differential

Proximate Traffic....................Filled White Diamond
- Within 6 Miles ***and***
- Within ± 1200' altitude

Traffic Advisory (TA) Intruder....Filled Amber Circle
- Predicted path of another airplane which will enter TCAS conflict airspace within 25-45 seconds
- Aural Warning ...."*TRAFFIC TRAFFIC*"

Resolution Advisory......................Filled Red Square
- Predicted path of another airplane which will enter TCAS conflict airspace within 20-30 seconds
- Threat is 25 seconds from closest point of approach
- Vertical avoidance maneuver displayed on IVSI
- If range is farther than current ND scale, **OFFSCALE** displayed on ND

Required flight crew response
- Within 5 seconds of initial RA
- Applying G-forces of ±.25 G
- This is more than AP can supply, so response must be hand-flown (*no change from previous TCAS policy*)

737 Briefings & Notes

Flight Attendant Briefing

Known delays
Maintenance—cabin or galley writeup and deferrals in AML
Ground—Short-taxi & safety demo considerations
Cabin Door Handles—Push Down (FMV1 Preflight 35.5.1)
Crew
Number of F/As and minimum crew requirements (OMV1 3g.10.2 Flight Attendants)
Coordination of crew meal service, if applicable
Time for meals is left to the CA's discretion
Need to minimize noise & activity near crew rest areas
Security
Details not included here for sensitivity concerns
Turbulence
Weather/turbulence forecast: duration & severity
Suspension of service at at flight attendant discretion
Notification required action for unexpected turbulence
Notify flightdeck of cabin situation & injuries if applicable
If not seated at jumpseat with interphone, stay seated to avoid injury until instructed to resume duties by Captain
International
Gen DECs/Customs Immigration forms (if required)
FOM 3c.4.2

TEST Briefing

T - **T**ype of emergency: general description of emergency.
E - **E**vacuation: Evacuation is anticipated.
Include any special considerations that may effect the use of exits.
S - **S**ignal: the following PA: "This is the captain. Evacuate. Evacuate. Evacuate."
T - **T**ime to landing.
QRH MISCELLANEOUS 0.45
FOM 1N.6.3 Evacuation

Unofficial Guide ** Training Only

— 1 —

Fold

ABC Approach Planning

A—ATIS
A—Approach Page
B—Box—Approach Loaded and checked
B—Brief Approach
C—Communicate--F/A (TEST) - PA - Company (7700)
C—Checklist

Holding Speeds

≤6,000........200 KIAS
>6,000 through 14,000........230 KIAS
May be restricted to 210 KIAS when depicted on a low altitude enroute chart;
Aircraft with an operational requirement to hold at airspeeds in excess of 210 KIAS must advise ATC prior to entering the holding pattern.
>14,000'........265 KIAS
FOM ARRIVAL 8g.3.3

Notes

Unofficial Guide ** Training Only

— 6 —

Cut ✂

Fold

Standardized Actions

Standardized Actions
Callouts are summarized in the table below: PF PM

Takeoff	T/O Engine Failure	2 Engine Go-Around	1 Engine Go-Around
Thrust Set, <N1> 80 Check V1 Rotate Positive Rate Gear up	Thrust Set, <N1> 80 Check V1 Rotate Positive Rate Gear up	Go-Around (TO/GA) Flaps ____ Positive Rate Gear up Set MAP altitude	Go-Around (TO/GA) Flaps 1 Positive Rate Gear up Set MAP altitude
Passing 400' AGL			
Select / Verify Roll Mode	Select / Verify Roll Mode	Select or Verify Roll Mode	Select or Verify Roll Mode
> 1000 AGL	E/O Accel Alt	> 1000 AGL	
VNAV or Set Speed (if VNAV N/A) Retract Flaps Climb @ UP Bug	VNAV or Set Speed (if VNAV N/A) Retract Flaps	Retract Flaps "Bug Up" is Automatic	VNAV or Set Speed (if VNAV N/A) Retract Flaps
> 2500 AGL	@ UP Bug		
VNAV (If not previously engaged)	LVL CHG, MCT ____ Checklist	VNAV or LVL CHG	LVL CHG, MCT ____ Checklist

References:
Summary Table........OM Volume 1, General 10-4
Takeoff........OM Volume 1, TAKEOFF 42-1
HUD Low Vis T/O...OM Vol. 1, TAKEOFF—HUD Low Visibility 43-1
Also other locations on profile pages
Takeoff Engine Failure........OM Volume 1, MANEUVERS 300-8
2 Engine Go-Around........OM Vol. 1, GO-AROUND 56-1
1 Engine Go-Around........OM Volume 1, MANEUVERS 300-14

Maximum Flap and Gear Speeds

1/2/5—240	25—190	<u>Gear</u> Ext—270/.82M
10—210	30—175	Extended—320/.82M
15—200	40—162	Retraction—235/(N/A)

Unofficial Guide ** Training Only

— 5 —

Operational Notes and Lists

Disclaimer**: On this page and the one following page are lists and consolidated information which have operational relevance. These lists are intended as a starting point for consolidating each pilot's personal notes, and the reproducible pages here are* ***For Training Purposes Only

This material is provided in an effort to help in consolidating policy guidelines for planning purposes. It is not an attempt to replace Operating Manual guidance from which these notes are derived.

Departure Briefing

Threats (PM then PF):
Relevant threats/concerns see:
—Potential Threats
—Ops Advisory pages (##-7)
Plan (PF):
—Taxi:
- planned route (including hot spots/runway crossings)
- departure runway

—takeoff:
- performance
- rejected takeoff considerations

—departure:
- planned departure (initial heading/altitude/fix)
- emergency (EO SID, EO acceleration/altitude, alternate, return)

OMV1 Preflight 35.5.2

Single Engine Taxi

Not authorized when:
Excessive taxi thrust required
Ramps & taxiways slippery
Braking action less than good
Icing conditions (unless no significant precip occurring)
APU INOP, except "maneuvering...for...X-Bleed start"
May not be advisable when:
MEL items involving normal / back-up modes of
◆Generators ◆Hydraulics ◆Steering & Brakes
Ramp or taxiway congestion, or
◆Tight turns ◆Uphill ◆Hazards
Proximity of the gate to departure / arrival runway
Safety compromise/efficiency of req'd checks, etc.
APU Use
Any taxi (SE or 2 Eng) OK with or without APU
If APU is shut down, can start remaining engine with X-Bleed Proc
OMV1 After Start 40.3.2

Unofficial Guide ** Training Only

— 2 —

Fold

Standard Thrust Not Permitted

The following require Maximum Takeoff Thrust but thrust ratings *22/24/26K* and *TO2/TO1/TO* may be used
TOW > ATOW (*Can get new TPS*)
MEL / CDL items
With weight corrections ***not*** on TPS (TPAS)
With MEL/CDL prohibitions of Std Thrust
Engine Anti-ice used; TPS not for A/I
TO data is from Contaminated Runway Correction (CRC)/ MEL message for an MEL/CDL item.
Maximum Thrust Required with *ONLY 26K*
Improved Performance
R/W contaminated-standing water, slush, snow, or ice
Windshear reported / expected / advisories; Use T/O flaps 5, 10 (-MAX 8 only), or 15 unless limited by obstacle clearance and / or climb limit
When FM-II Airport Advisory requires Max
27K Bump Thrust
Always requires the use of max thrust
Standard thrust is prohibited.
PERF-TPS 10.4; PERF B737 Takeoff Card

Maximum Thrust 26K (737-800), or TO max (737-MAX)
Recommended for
Gusty winds (>10 Knot gust factor)
Strong Crosswind (>15 Knots crosswind component)
OMV1 Takeoff 42.4.5, LIM 10.10

Approach Setup—WARM-V

Initial Setup—"WAARM-V"
<u>W</u>eather requirements & NOTAMS—Check
Include crosswinds, visibility, published minimums
<u>A</u>pproach—Load Approach from Database
Brief Approach
<u>ANP</u>—If RNP<0.3, check Dispatch ANP valid time period
<u>R</u>NP—Check chart & Set As Required for the Approach *
<u>M</u>CP—When cleared for the approach—LNAV / VNAV
Set FAF altitude when on intercept or a portion of the approach *
<u>V</u>OR Updating—Turn Off (<PROG> <NAV DATA>) *

*NON-ILS

Unofficial Guide ** Training Only

— 3 —

Cut ✂

Fold

Passing FAF Actions

Passing FAF Actions—MAC-V
<u>M</u>issed Approach Altitude—Set when 300' below MAP
<u>A</u>ltimeters—Verify CA—F/O + 100' *
<u>C</u>learance—Tower Call Passing FAF
<u>V/S</u>—Set prior to MDA if end point not RWXX *
*NON-ILS

Final Approach Deviation Callouts

Airspeed..+10 / -5 knots*
Rate of Descent.......................<2000'—More than 2000 fpm
.......................<1000'—More than 1000 fpm*
Glide Slope...> ½ Dot Deviation
Localizer> ....⅓ Dot (non expanded) or
.........Diamond touching Rectangle (expanded)
Touchdown....First 1/3 of Runway Not to Exceed first 3000' *
PNF directs GO AROUND at 1000'AFE in IMC or 500'AFE in VMC if stabilized approach criteria are not met or maintained throughout remainder of approach.
FMP1 APPROACH 51.4.5 to 51.4.10

AutoBrakes

Must be armed (If operative) for any of the following:
Runway — Wet or contaminated or
Less than 7000' (6000' or less requires 3 or MAX)
Weather — Tailwind or
RVR less than 4000 or visibility less than 3/4 mile or
Braking conditions reported less than good
Recommended: landing with gusty winds or crosswinds
OMV1, LANDING 55.6.2

Flaps 40 Landing Recommended

Braking action < than good
Tailwind
Wet / Contaminated runway
Runway <7,000'
OMV1 APPROACH 51.1.2
OMV1, LANDING 55.1.3

Unofficial Guide ** Training Only

— 4 —

Titles Currently Available

By Rick Townsend

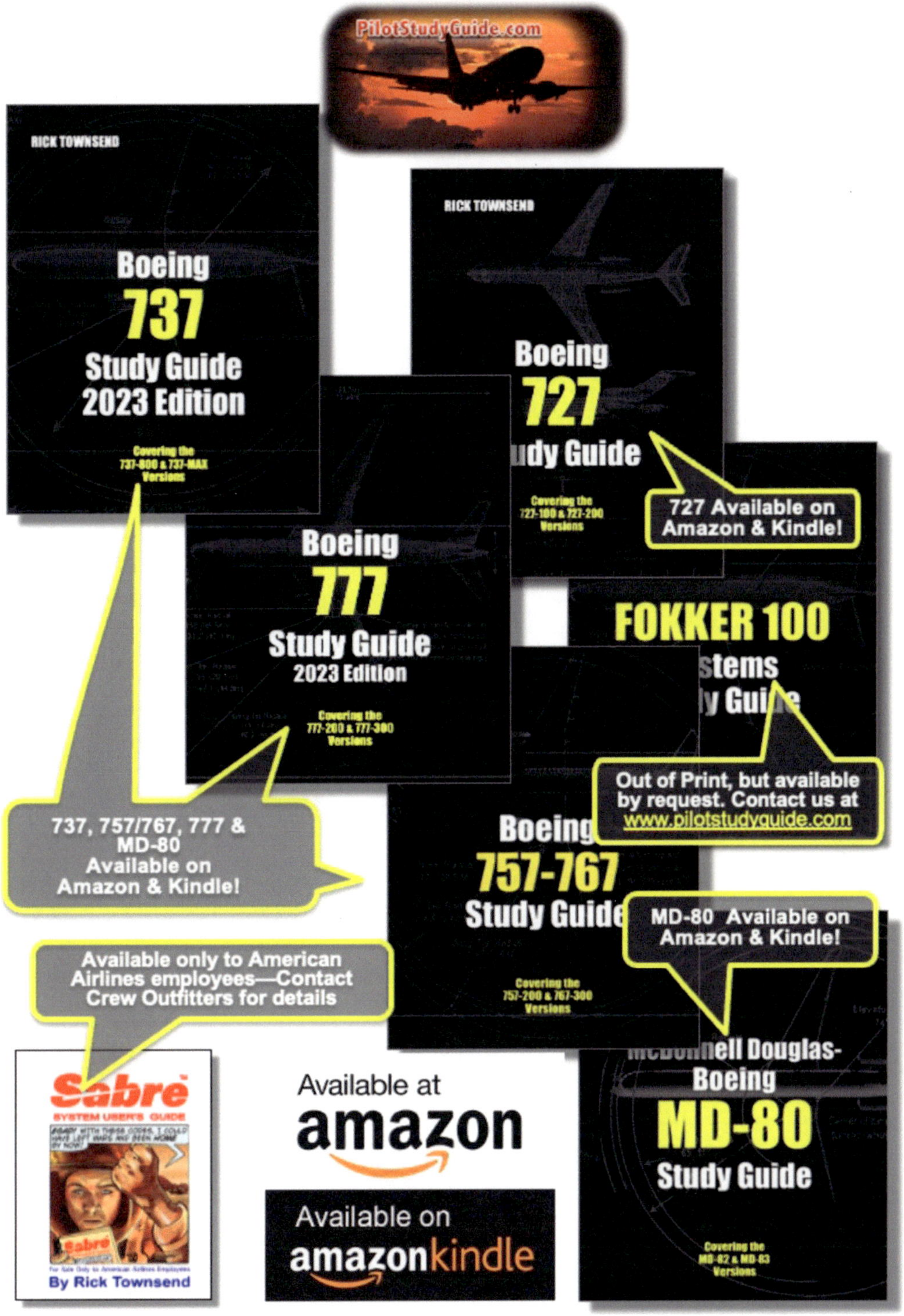

For more information on these books, see our web site for Pilot Study Guides

To purchase any of these titles, go to Crew Outfitters locations at the Flight Academy Gift Shop, or at DFW Airport Terminal C Operations
To purchase online, go to *www.crewoutfitters.com*

To order by phone, call **800-874-1204**

This book is available through Amazon in both Kindle and in print format!

Giving Back

A portion of the proceeds from every book help support some worthy charities. Here are two of the charities your book purchase supports:

Kenya Kids Can!

Feeding and teaching world changers of tomorrow

Kenya Kids Can is a non-profit working in the Rift Valley of Kenya. They feed students and construct solar powered computer centers to improve education and bring hope to needy students. Please browse the website to learn about what can be done to truly change lives.

www.kenyakidscan.org

Snowball Express

GARY SINISE FOUNDATION · SNOWBALL EXPRESS · PRESENTING SPONSOR American Airlines

Serving the *children* of our fallen military heroes

Snowball Express Could Use Your Help!

Snowball Express honors America's fallen military service members who have made the ultimate sacrifice since 9/11 by

- **Humbly serving the families they left behind**
- **Championing their children's future success by creating opportunities for joy, friendship, and communal healing**
- **Connecting these families to one another.**

Since 2006, the mission of Snowball Express has been a simple, yet profoundly important one: Providing hope and new happy memories to the children of military fallen heroes who have died while on active duty since 9/11. In December each year we bring children together from all over the world for a four-day experience filled with fun activities, like sporting events, dances, amusement parks and more.

Nationally, Snowball Express provides comprehensive support programs for fallen families that are focused on transition and connections to community resources, healing and wellness, peer engagement, education and personal/professional development programs.

www.snowballexpress.org

Made in the USA
Monee, IL
12 September 2026

62974521R00045